Quantum Computing Secrets

The Ultimate Human Guide to Quantum Universe

Quantum Physics and Quantum Mechanics Supremacy Explained For Beginners And Their Impact on AI, Finance, Crypto... And Your Daily Life

By Laurent Meri

And

Carl Prometheus

Quantum Computing Secrets

Carl Prometheus and Laurent Meri

Published by LAURENT MERI, 2023.

While every precaution has been taken in the preparation of this book, the publisher assumes no responsibility for errors or omissions, or for damages resulting from the use of the information contained herein.

QUANTUM COMPUTING SECRETS

First edition. August 25, 2023.

Copyright © 2023 Carl Prometheus and Laurent Meri.

ISBN: 979-8223137009

Written by Carl Prometheus and Laurent Meri.

Table of Contents

Wanna Hear a Little Secret? Keep It Short

In today's world of super-short attention spans, you gotta trim the fat if you want to pay attention to hot takes. Sure, jumbo books look impressive on your shelf. But when it comes to real talk, shorter is sweeter.

Concise books give you a laser-focused highlight without all the boring backstory or fluff. Their every sentence packs a punch, not wasted words.

Got an earth-shattering idea? Don't bury it under 600 pages of dust-dry prose. Distil that sucker down to its purest, minimalist form. Like a perfectly blended smoothie, filter out the unnecessary chunks for maximum impact.

Besides, who has time for lengthy door topper tomes anymore? We want knowledge in bite-size nuggets now. With fit-for-purpose books, you can suck down those nourishing morsels in a few sittings, retaining way more. Really let the savoury flavours marinate.

And don't forget the sharing! Itty bitty ideas spread like hot fire because everyone has time for a quick but amazing story. Just ask that monk dude with the Ferrari - his cliff's notes wisdom went viral faster than a Kanye rant.

So do yourself a favour and keep it short. Chop off any unnecessary padding. Get straight to the good stuff and deliver it with gravitas.

Less is more when you choose each word with care.

Introduction

What up home slices, and welcome to our quantum party extravaganza! Buckle up buttercup, because we're about to take an epic joyride through the wildly weird yet massively powerful world of quantum computing.

I'm talking strange science that would make Einstein and Schrödinger trip on their own brains. This stuff gets whacky! But stick with me amigos, and you'll be dropping quantum knowledge bombs at parties in no time.

See, that's my mission here - to make quantum computing approachable for regular folks, not just PhD robot nerds. We're all invited to get quantum curious! I won't leave you in the dust with any wacky equations or technical nonsense.

We'll explore how quantum works in simple enough terms to pump up your knowledge without needing an MIT degree. Though if you happen to have a particle physics doctorate, no judgement! We accept all breeds.

Now I bet you're wondering, what the heck makes quantum computing so radical compared to the iPhones and laptops we know today? Great question!

The big difference is that today's devices operate using good old fashioned binary bits - those basic units of data that are either 1 or 0. But quantum takes things to infinity and beyond with its crazy qubits!

Instead of just being 1 or 0, qubits can blend into this mysterious in-between state called superposition. It's like 0 and 1 get freaky and produce a quantum baby!

Suddenly one qubit isn't limited to just two states - when you link them up, you get an endless multiverse of possibilities and parallel calculations all occurring simultaneously. NBD.

It's like turning your Xbox into a time-traveling teleportation device that predicts alternate futures. Today's computers get drunk just thinking about quantum's potential!

These wildcard qubits don't just stop at individual weirdness either. Thanks to the magic of entanglement, qubits can also stretch their spooky connection across vast distances.

So two particles on opposite sides of the galaxy can influence each other instantaneously as if they're telepathically linked. You think 5G is fast? Quantum laughs at our human notions of speed.

When you combine all these wild abilities for exponential parallelism, superpositions, and telepathic entanglement, you open the door to computing power we can barely even fathom with our limited lizard brains.

It's like we just discovered an ancient book of spells for summoning transdimensional demons to do our computational bidding. Things are about to get crazy up in here!

Of course, we have to master the hocus pocus first. Taming quantum's exotic potential into useful applications is still a major challenge.

Qubits tend to "decohere" (science talk for freaking out and forgetting what they're doing) when we try to interact with them. It's like trying to get a toddler hopped up on Skittles to focus. Madness!

But quantum researchers are quickly conquering these roadblocks using tricks like error correction and noise-reducing algorithms. Not gonna lie, it's still early days.

However, we've recently seen the first big quantum milestones that provide a glimpse of just how revolutionary this tech could be when fully controlled.

For example, Google achieved quantum supremacy in 2019 by using a 54-qubit system to perform a calculation in 3 minutes that would take 10,000 years on a classical supercomputer. Bonkers!

With that taste of the quantum potential, many experts predict we're less than a decade away from having quantum computers capable of breaking all previous limits in fields like chemistry, AI, cryptography, finance and more.

That's what makes now the perfect time to wrap your head around quantum's endless possibilities! We want to be like Marty McFly cruising around in the DeLorean, not baffled cavemen encountering a smart phone for the first time.

Through this book, you'll gain a solid understanding of how quantum works so you're prepared for the coming computing revolution. Consider it like a crash course in Quantum 101 - no confusing math or padding, just the highlights distilled down.

We'll demystify qubits, entanglement, and how quantum algorithms like Shor's and Grover's provide such an exponential advantage. You'll also get up to speed on the key players and technologies racing to unleash this radical power.

By the end, you'll have cultivated that crucial quantum intuition to envision the seismic changes coming as these technologies mature in the next decade. The future is gonna get weird!

So buckle up and get pumped, curious reader. We're heading on an epic yet digestible adventure together to unlock the mysteries of the quantum realm! No PhD required.

6

This is our chance to peek behind the curtain at this emerging new paradigm of computing with galaxy brain potential. Let's dive in! The quantum multiverse awaits...

A Note Before We Dive Into The Quantum Realm...

First, massive thanks for picking up this book and joining the quantum curiosity club! By deciding to level up your quantum knowledge, you're boldly going where few have gone before. Major props.

Now I have a huge favour to request. If you find this beginner's guide useful, it would be cosmic if you could take just a minute to leave an honest positive review on the platform you bought it from. Let me explain why it matters so much:

Positive word-of-mouth is essential for niche topics like quantum to spread. Your review gives the algorithms a signal that these concepts are resonating with real humans (aka you!).

This then allows me to reach and enrich many more curious minds out there. My dream is igniting a passion for science at scale and making quantum accessible!

But that can't happen without reviews from stellar readers exactly like you. Seriously, you da best. So if this book clicks with your quantum cravings, please boost the signal with a positive review. It would mean multiple universes to me!

Quantum computing needs all the buzz we can generate. With your small action, together we can compound interest in this field and inspire the next generation of pioneers.

Plus, you'll likely motivate someone you know who needs this quantum primer too. Spreading the love sparks a chain reaction.

So truly, thank you for even considering leaving your feedback. Now let's immerse into the mystical quantum multiverse! Ad astra!

Chapter 1 - Quantum Computing Basics

What Makes Quantum So Special?

Qubits are radical - they allow a crazy explosion of parallel universes!

Buckle up buttercup, because we're diving deep into why qubits are the key to unlocking quantum computing's insane parallel processing powers! Grab your photon packs and let's get quantuming.

First up - qubits operate according to the wild and wacky laws of quantum mechanics, which differ radically from the common sense rules we're used to in our human-scale reality. These quantum traits open up radical new possibilities!

For example, at super small subatomic scales, particles can actually exist in an uncertain hazy state of all possible values until they are measured and forced to pick one. Say what now?

It sounds bizarre, but this probabilistic fuzziness is a key quantum trait that gives qubits their extraordinary capabilities once harnessed. Let's break it down...

In non-quantum computers, the basic unit of information is the trusty binary bit, which can only be in one of two definite states - either 0 or 1. Pretty boring and limited.

But quantum's qubits blow the doors wide open! That's because qubits can exist in a superposition - meaning they can embody the properties of 0 and 1 simultaneously before observation. Mind blown!

It's like qubits give zero and one the fused powers of Dragon Ball Z's Goku and Vegeta all rolled into one microscopic package. Their powers combine to transcend binary limits!

Now you may be thinking - hold up, how can a qubit possibly be both 0 and 1 at the same time?? That seems totally illogical and nonsensical based on reality as we know it! An excellent point, dear reader.

Here's the trick - at incredibly tiny subatomic scales, particles can actually exist in an uncertain probabilistic state of all values until measured. Schrödinger's cat being dead and alive and all that jazz.

So a qubit remains in an ambiguous superposition until you take a measurement, which causes it to randomly collapse into being either 0 or 1. Wild, right?

Let's break this down with a thought experiment. Imagine a single qubit as a coin flip in mid-air...

As that coin spins through the quantum realm, it exists in a simultaneous state of both heads and tails - akin to 0 and 1 - until an observation forces it to pick a side and collapse into a definite state upon landing.

In the blinking of an eye, a qubit can flip between embracing all possible values to settling on just one. It's like flipping a coin and having it come up both heads AND tails, until you look! (Spooky action at a distance.)

And here's where qubits really blow up - when you chain them together in large numbers, the possibilities explode exponentially thanks to those uncertain superpositions.

With just 50 entangled qubits, you unlock over a quadrillion potential parallel states! Compared to a classical bit string which has only a single definite combination.

This means a quantum computer can represent and rapidly calculate across all possible solutions at the same time in parallel dimensions. It makes today's computers look dumber than your average sloth.

We're talking answering complex questions in minutes that would take today's supercomputers longer than the age of the universe. Now that's what I call a glow up!

Here's a simplified real-world example to make this hit home...

Let's say a bank wants to analyze historical stock market data to find the absolute optimal portfolio that would have produced the highest possible returns in the past.

Well with 10 stocks to choose from, a classical computer would need to methodically test all possible combinations of assets to invest in - and with 10 stocks, that's over 1000 combinations to individually analyze.

But a quantum computer can just represent a superposition of ALL those various portfolios at once - then instantly perform calculations on them in parallel universes to spit out the single best choice! It's over before it even started.

We go from ploddingly testing options one-by-one to instantaneously visualizing the entire solution space and evaluating it simultaneously. Thank you, strange quantum magic!

"But wait!" You may be thinking, "Doesn't this violate the supposed impossibility of time travel and knowing the future? Have we broken causality itself??"

An excellent observation! And no, we haven't enabled actual clairvoyance quite yet. Here's the catch...

Once a quantum calculation ends, the qubit superpositions instantly collapse back down to a single definite state.

So in our investing example, we'd see the ideal historical portfolio but can't actually change the past. The quantum wavefunction returns to solid reality.

You can think of it like this - superpositions give us a temporary glimpse into the Platonic realm of potentiality outside time and space, represented by equations.

But then qubits collapse back into singular experimental results grounded in the here and now. Wild stuff!

The key is that superpositions allow efficiently probing WHAT WOULD have happened in all parallel worlds and histories, without the ability to alter the past or transcend time itself. Bullet dodged!

Alright, let's take a deep breath because I know your brain is currently overflowing with possibility and burning questions at the moment. Totally normal!

For now, let's just be content knowing qubits give us a computational peek behind the cosmic curtains to realities beyond time and space thanks to quantum superpositions. The limits are gone baby!

There's much more quantum mystical magic to come. But I hope this section helped explain why qubits are so radical and pivotal to unlocking unbelievable quantum capabilities through probable parallel universes!

Entanglement and superposition create freaky particle connections across cosmic distances!

Alright, we've explored how superpositions give individual qubits radical uncertainty powers. Now let's discuss how entanglement creates even freakier connections between pairs of qubits across vast distances!

With superposition, one qubit can represent multiple values at once in a probabilistic quantum state before observation collapses it down to a single possibility. Pretty nutty already.

But entanglement takes it up a notch by generating spooky instant correlations between two particles that essentially fuse them into a single quantum entity - even when separated by distance!

Let's unpack this...

When two qubits become entangled, their states and properties become deeply interdependent regardless of physical proximity. Their fates are eternally intertwined!

It's like entangling your heart with your crush's heart. Suddenly when her heartbeat flutters from excitement, yours does too even miles away as if magically linked. Awww...

But entanglement's correlations are actually much wilder than teen romance. Get ready for some next level action!

Picture this: we take an atom and split it, emitting two entangled electrons. Quantum cuckoo clock.

We separate these electrons to opposite ends of the observable universe - billions of lightyears apart. No big deal.

Now, when we measure the spin of electron A on Earth, we immediately know electron B's correlated spin a bajillion miles away will be in the opposite direction, faster than you can say "spooky action at a distance"!

It's as if these particles communicate telepathically to instantaneously coordinate properties across cosmic distances at speeds exceeding light's supposed cosmic speed limit. (Take that, Einstein!)

In fact, this is where the famous EPR quantum paradox comes from that stumped even the greatest minds like Einstein - how can entangled particles remain so freakishly linked when separated by space and time? Seems impossible based on classical physics!

Of course, the problem is that we're applying common sense human-scale logic to the utterly absurd and fantastic quantum domain - a place that laughs in the face of our limited conceptions of causality and concrete reality. Oops.

But the fact remains - experiment after experiment conclusively confirms entangled particles somehow remain connected in this mystical quantum way that transcends conventional spacetime bounds.

We're talking instant, perfect coordination of states between particles lightyears apart - with zero lag or discernible communication mechanism. Space and time melt away!

Now, we do have to be careful not to get too loopy and assume entanglement will let us communicate faster than light. Information still can't be transmitted this way.

But the correlations are real. Particles become quantum conjoined twins splitting apart yet remaining eerily unified across any distance.

So in summary, quantum entanglement leverages the same probabilistic fuzzy logic that powers superposition, but shared between particle pairs instead of individuals.

And when you chain together many entangled qubits in a quantum computer, the exponential parallel possibilities and instant correlations continue multiplying.

Suddenly a quantum machine doesn't just calculate probablistic superpositions internally. It also instantly coordinates complex probabilities across all its entangled qubits!

This allows exponentially swelling amounts of information to remain in perfect harmony within the computer architecture as it performs massively parallel operations. Jaws on the floor.

Let's step back to look at why this matters in bigger picture terms:

In a classical computer, each additional bit doubles the information capacity. 1 bit holds 1 value. 2 bits hold 2 values. 3 bits hold 4 values. And so on.

But with quantum entanglement, adding more entangled qubits doubles the possibilities exponentially! So 100 qubits entangled together gain 2^{100} values - a number so mind-bogglingly massive that all the atoms in the universe could not store it classically!

This endless information scaleability is what truly unlocks unbelievable computational power. Entanglement alongside superposition is the peanut butter and jelly that makes the quantum sandwich shine.

Alright, let's take a breather because I'm sure your mind is reeling from envisioning these bizarre entanglement connections transcending space and time. Freaky stuff!

The key takeaway for now is that entanglement Allows pairs of particles to become linked as eerie quantum twins with instantly coordinated properties regardless of physical separation. Wacky, but experimentally confirmed!

Harnessing this quantum mysticism allows the information capacity and parallel coordination inside a quantum computer to scale exponentially towards infinity. And that's what's gonna take computing to the next level!

Okay, ready for more quantum weirdness? Let's keep exploring additional unique traits that enable these insane abilities. The quantum joyride continues!

Quantum computers calculate all the possibilities at once. They make today's computers look like chumps!

Alright, we've covered how individual superpositions and spooky entanglement connections give qubits their radical uncertainty powers. Now let's discuss how this enables calculating all possibilities simultaneously - leaving old school computers choking on quantum's dust!

Thanks to superpositions, a single qubit can represent multiple ambiguous states at once in a probability wave before observation forces it to collapse into one.

Meanwhile, entanglement allows this probabilistic coordination to occur across many qubits simultaneously, regardless of physical proximity.

Together, these attributes allow quantum computers to instantly calculate across all potential options and parallel universes! Jaws on the floor.

Let's understand this through a simple example comparing classical and quantum calculations...

Say we want to know the factors that make up a large number - like what are all the possible pairs of numbers that multiply together to get 56?

Well a normal computer would start dividing 56 by factors in a linear sequence until it found the answer: 2 x 28.

This step-by-step testing of divisors in order wastes tons of time, especially for massive number factoring problems like decoding encryptions.

But a quantum computer can instantly represent ALL potential factors for 56 in a superposition - meaning it tests every option simultaneously in parallel universes!

We go from ploddingly analyzing one divisor at a time to instantly envisioning the entire solution space and querying it all at once. Huge difference!

This allows a quantum machine to derive the same result exponentially faster by probing a probabilistic realm of all possibilities outside normal time and space. VR hackers!

Let's step back and compare how classical bits vs qubits enable this:

A classical bit string moving sequentially through combinations suffers from a massive bottleneck. It must trudge through options step-by-step in slow human time.

But when you chain together hundreds of ambiguous qubits entangled in superposition, you can effectively compute across quadrillions of parallel states and histories simultaneously!

We're talking answering in minutes questions that would take today's supercomputers longer than the age of the universe. Your smartphone just cried itself to sleep.

Here's another real-world example:

Say you want to analyze historical weather data to find patterns that predict the optimal time and place to install solar panels for maximum energy generation.

Well with hundreds of location variables and terabytes of atmospheric data, testing every possible combination on a normal computer would be hellishly slow.

But throw that dataset into a quantum system - and suddenly it can instantaneously project and analyze across parallel superpositions of ALL possible weather histories and solar configurations at once!

This reveals the ideal solar panel solution without having to slowly grind through every option step-by-step. The quantum calculation finishes before it even started. A true Skip to the Good Part machine!

And it's not just solar - this works for chemistry simulations, financial predictions, machine learning, cryptography and more. Any complex multivariate problem, quantum can solve before you blink.

That's because qubits allow computationally probing a platonic hyperspace representing all possibilities beyond normal time and space. Talk about hacking the matrix!

Of course, we do have to collapse the quantum wavefunction back down into one definite answer we can measure in classical reality once the processing finishes. No actual time travel here!

But temporarily glimpsing all probable paths forward enables solving problems intractable to classical systems. That's the real power!

Alright, let's take a breath because I'm sure envisioning instantaneous multiverse computations makes your head spin like a drunk ballerina. Mine too!

But the core idea is: superpositions + entanglement = calculating across parallel probabilities simultaneously = mindblowing speed and efficiency improvements.

By transcending classical sequential processing limits, quantum lets us probe hypothetical what-if scenarios spanning endless options in mere minutes rather than eons.

So in summary, quantum parallelism leaves conventional silicon computers in the dust. This lets us solve previously unsolvable problems and envision revolutionary possibilities!

Alright, ready to dive back into the quantum zone? Let's continue navigating this weird and wonderful physics enabling unbelievable future applications. The adventure continues!

Comparing Classical and Quantum Bits

It's like zero and one had a baby named INFINITY.

Alright, now that we've explored the mystical traits of qubits, let's compare them to classical bits to truly appreciate the exponential difference. Time for a good old fashioned quantum smackdown!

In traditional computers, the basic unit of information is the classic binary bit - represented by either 1 or 0. On or off. These definite states store data and perform logical operations.

But quantum computing unlocks radical new possibilities by introducing uncertain qubits - which can exist as 1 AND 0 simultaneously before observation, thanks to superposition.

It's like qubits give zero and one the fused powers of Dragon Ball Z's Goku and Vegeta - combining their elements to reach a level of infinity. Their quantum baby breaks all limits!

Let's visualize this difference. Imagine a classic bit is like a coin - it's either heads or tails. Definite states. Pretty basic.

But a qubit is more like Schrodinger's cat - mysteriously both dead and alive until you open the box and force a measured state. The qubit embraces all potentialities.

This means a single qubit can represent a superposition of 0 AND 1 at the same time - allowing computation across both values simultaneously versus just one binary state. Huge upgrade!

To put it in math terms, a classic bit gives 2 finite options: 0 or 1. But a qubit provides a vector of probabilities between 0 and 1 with endless configurations. We've quantum leapt!

When you chain together many qubits, their probabilistic superpositions multiply exponentially towards infinity. We get 2^100 combinations instead of just 100 binary strings.

This exponential scaleability is what enables quantum computers to solve problems intractable to classical systems - by harnessing probability waves in quantum space.

Let's step back and compare how these different units process information:

Classical bits must move through combinations sequentially in plodding human time. Testing options one by one leads to a huge bottleneck. Like a donkey walking a tightrope.

But superposed qubits allow simultaneously computing across all options in parallel universes outside time and space! We probe the platonic realm of potentials.

It's like pixels loading an image line by line versus having the full picture instantly. Qubits give the complete vision!

Here's a more tangible example...

Say Google wants help deciphering complex Chinese text into English. Well serially testing translation combinations on a normal computer would take forever.

But their quantum computer can simultaneously represent a superposition of ALL possible English translations at once in parallel, allowing instant pattern matching to identify the right choice!

So while a classical machine must slog through each variant slowly, a quantum device simply checks all options immediately like Neo visualizing the Matrix. Infinity baby!

This works for chemistry simulations, financial predictions, cryptography and more. Any multivariate problem, quantum cracks before you blink.

That's because qubits allow computationally exploring a hypothetical multiverse of possibilities beyond normal time and space constraints. Talk about life hacks!

And thanks to entanglement, probabilities scale exponentially with more qubits since their values stay coordinated across space. 100 qubits gives 2^100 values!

Of course, at the end of the calculation we still have to collapse the wavefunction down to one state we can measure in classical reality. No actual time travel here!

But being able to efficiently probe hypothetical what-if branches allows solving problems intractable to step-by-step binary processing. That's the key quantum edge.

In summary, qubits' uncertain superposition states unlock the truly exponential scaleability needed to exceed classical computing constraints. It's like bits got pumped up on quantum HGH!

So while traditional bits can only be definitively 0 or 1, qubits embrace probabilistic infinity - and that's what opens the door to unimaginable new computing potential.

Alright, let's take a breath and let this epic beatdown between classical and quantum sink in. I think we all know who emerged victorious! Mic drop

But in seriousness, I hope this section helped crystallize why qubits are such a monumental departure from traditional binary bits. They bring infinity to computing!

Translating this science mumbo jumbo into simple terms

Alright, now that we've explored how radically different qubits are from classical bits, let's make sure all this crazy quantum stuff is breaking down into understandable concepts. I want your quantum knowledge turned up to 11 by the end!

I know - superpositions, entanglement, probability waves, platonic hyperspace, collapse of wavefunctions. It's a lot of science mumbo jumbo!

But stick with me, and we'll boil these concepts down into digestible quantum nuggets of wisdom. Let's do this!

First up: superpositions. This just means a qubit can represent multiple potential states simultaneously before observation "collapses" it into one option, thanks to quantum uncertainty.

Remember our quantum coin flip analogy? The qubit is metaphorically the coin spinning through all values of heads and tails before landing in a definite state when measured.

So superpositions simply refer to qubits embracing all probabilities at once in a mysterious quantum limbo. We can make this tangible by comparing it to familiar things:

A superposed qubit is like Schrodinger's cat - simultaneously alive and dead until observed. Or like standing at a crossroad with all paths ahead.

In math terms, it means inhabiting all points on a graph before picking one. In decision terms, envisaging all options before choosing. Open potential!

Superposition is the source of qubits' power. Unlike rigid binary bits, they represent infinite freedom and parallelism.

Now onto entanglement - the ability for qubits' probabilities to stay weirdly interlinked regardless of physical distance. Einstein called this "spooky action at a distance".

Let's break this down into something more relatable:

Think of entanglement like best friends who can sense when something is up with the other, even miles apart. Sympathetic vibrations!

Or imagine particle pairs as identical twin qubits with innately correlated properties even when separated. Built-in harmony across space and time.

In practical examples, this allows entangled qubits in a quantum computer to act as one entity. Their values stay unified no matter the physical gap between them.

So in summary, qubit entanglement means innate connection at a distance. Freaky quantum friendships!

Alright, probability wavefunctions just refer to the ambiguous state qubits inhabit before observation forces a concrete reality.

Imagine it like a cloud of possibilities that temporarily allows occupying all options, before precipitating into one definite outcome when measured. Quantum rain!

And finally, collapsing the wavefunction simply means taking that measurement that forces superposed qubits to pick one state.

It's like opening the box to look at Schrodinger's cat - suddenly it's either alive or dead, no longer both. Our observation crystallized reality.

Phew! As you can see, even complex quantum traits become intuitive when we use familiar analogies as cognitive stepping stones. Don't let the jargon intimidate you!

At its core, quantum mechanics simply embraces possibilities and freedom instead of rigid binaries. It's like upgrading from a flip phone to an iPhone with more computing power.

Harnessing uncertainty may seem bizarre at first. But these quantum operating principles are mathematically sound and technological progress is making them highly practical.

So while the inner workings of quantum computers may seem otherworldly, they're built on solid science and represent the coming evolution of computing.

The key is having an open and curious mind. Don't get discouraged when quantum concepts seem confusing or counterintuitive at first glance. Stick with it!

Let your intuition expand slowly to encompass these new possibilities. With time and practice, you'll develop that gut quantum sense just like the pioneering scientists.

Alright, let's take a break here because I know your brain is likely full to bursting! Don't worry, we'll continue this epic quantum knowledge journey together step by step.

For now, I hope this section provided a friendlier introduction to some of the core quantum concepts powering this computational revolution. Let me know if any part still feels unclear!

You got this. Just remember, at its heart, quantum mechanics is about embracing freedom, infinite possibilities, and all the complexity of our universe. A truly awesome upgrade!

Okay, ready to continue diving deeper into the quantum realm? Grab your photon pack and let's go explore more of this intriguing new paradigm. The adventure continues!

Quantum Properties - The Weird Stuff

Tunneling, teleportation, interference, and more... the totally bizarre traits that give quantum computers their insane power.

Alright, we've explored some of the major quantum phenomenons like superposition and entanglement. Now let's unravel even more of the weird and wacky traits that contribute to quantum computing's otherworldly potential. Enter The Twilight Zone...

First up - quantum tunneling. This is the ability for particles to do the seemingly impossible - overcome huge energy barriers or even pass through impenetrable walls via quantum teleportation. Let me explain...

Classically speaking, if you were kicking a soccer ball against a massive concrete wall, no way could that ball penetrate the barrier without immense force. It's just common sense based on Newton's laws.

Ah but in the quantum realm, particles can borrow energy from the universe to channel Star Trek's Scotty and magically teleport to the other side of obstructions like it's no big deal! Qubits laugh in the face of walls.

It's like instead of futilely banging against the obstacle, the particle simply borrows enough energy to briefly turn into a waveform and zap through to the other side unimpeded. Ghost ball goes brrr.

In quantum tunneling, it's as if particles can tap into cosmic cheat codes and no-clip through boundaries that would stop other objects cold. The limits of classical physics don't apply!

Researchers have verified this funky phenomenon experimentally. Subatomic particles consistently exhibit behavior that should be impossible based on Newtonian premises. Welcome to the quantum zone!

Now, don't get too loopy here - macro objects like soccer balls still can't suddenly blink through walls. But this opens radical possibilities as we harness quantum principles.

For example, by leveraging quantum tunneling, future computing chips could allow electricity to flow efficiently even when blocked by supposedly impassable insulating barriers. Quantum pixie dust!

Alright, next up: quantum teleportation! Now before you get too excited, no, we haven't yet achieved Star Trek style matter teleportation beaming humans around. That would be pretty rad though.

But what we CAN do is use entanglement tricks to teleport quantum information instantaneously - including the precise state of qubits! - from one location to another without traversing the physical space between. Trippy!

This works because of entanglement's spooky action at a distance effect, which Einstein called "spukhafte Fernwirkung" - German for creepy ghostly remote influencing. Thanks Al!

Basically, when particle pairs are entangled, whatever happens to one impacts the other simultaneously, even light years apart. This creates a channel to transmit quantum data.

So if we entangle two qubits, then manipulate one to encode information, we instantaneously transmit that encrypted signal to the other qubit, no matter how far apart they are! Space and time, we come for you!

Again, we aren't literally teleporting matter Star Trek style yet. But we can harness entanglement to zap quantum information around with no lag, which has huge practical use cases.

For example, quantum teleportation could enable unhackable channels for transmitting classified data and encryption keys, since any tampering breaks the entanglement connection. Beam me up!

Alright, next on the quantum oddities list - wave-particle duality! This one blows minds...

You may have heard that subatomic entities can behave like both particles AND waves. It's weird because in our normal reality, things usually act as either defined particles OR waves, not both.

But in the quantum realm, photons and other entities regularly flip flop between seemingly contradictory states. It's like an identity and existential crisis!

When not being observed, quanta act like diffuse probability waveforms - vibrating in all locations simultaneously. But look closely and they suddenly snap into defined particles with set locations! What is this sorcery?

It's like trying to pin down a shape-shifter who keeps morphing between animal and human form. Our limited brains can't comprehend this fluid both/and existence.

But particle-wave duality is critical to the probability magic of quantum computers. Qubits can represent floating potentials before collapsing into defined states when measured. Their waveform nature enables uncertainty and superpositions.

So don't get deterred by how bizarre and contradictory this seems at first glance. Just step back and embrace how quantum expands our conception of reality's possibilities!

Okay, one more quantum quirk worth mentioning - quantum decoherence! This is essentially noise and interference disrupting those delicate quantum states.

You can imagine it like a fragile soap bubble - the slightest breeze will make it pop and collapse. Similarly the smallest environmental vibration decoheres qubits before we want them to settle into defined values prematurely. Not cool.

Quantum computers need to live in an extremely finely tuned environment for their fragile probability states to persist intact. Otherwise it's like trying to meditate during an earthquake! Concentration shattered.

But researchers are making progress with error correction techniques that help reinforce entanglement and allow qubits to maintain coherence stable enough for computation. Baby steps to quantum practicality!

Alright, let's pause here because I'm sure these concepts are pummelling your brain like a sledgehammer of what-the-physics right about now! Take a breath and don't worry.

The key takeaway is that quantum embraces all sorts of trippy properties that transcend rigid human concepts of definite realities. It's like graduating from preschool level to utopian singularity level physics!

Mystifying but mathematically sound. As we engineer ways to harness quantum's advantages, unbelievable new directions open up. Pretty cool right? Alright, ready for more spooky science magic? Let's go!

You're about to enter The Twilight Zone...dun dun dun!

Alright, we've explored some of the strange traits that give quantum computers their radical edge. Now let's dive even deeper down the rabbit hole into the utterly bizarre quantum realm! Cue the Twilight Zone music...

See, the deeper you look into quantum physics, the more it reveals mind-bending realities that seem impossible through our normal human logic and assumptions. It's like peering behind the curtain of reality itself!

At the smallest subatomic scale, our common sense classical physics completely breaks down. Cause and effect, space and time, identity and locality - nothing works like we expect. It's topsy turvy land!

Let me paint you a picture of some of the paradoxical quantum wonderland we're plunging into...

Schrodinger's Cat - Dead and Alive?!

Our first exhibit is Schrodinger's famous thought experiment. Imagine a cat locked in a box with radioactive material that has a 50/50 chance of decaying, which would release poison and kill the cat.

But until we open the box, we have no way of knowing the decay status. So that cat exists in a superposition of both dead AND alive states simultaneously - an utterly bizarre but valid truth in the quantum realm!

Only our observation forces the cat to "choose" alive or dead. But left unmeasured, it embraces the contradictory duality. Wild!

Spooky Action at a Distance

Next up - Einstein's spooky "spukhafte Fernwirkung." This refers to how entangled particles coordinate their properties FAR faster than light speed across vast distances.

Change one particle here on Earth, and its entangled twin instantly aligns its spin a galaxy away. But nothing seems to communicate between them! Mystical remote influencing defies spacetime.

Einstein hated this "too spooky" implication. But experiments conclusively verify this bipartite coordination happens instantaneously - leaving our human intuitions bewildered.

Warp Speed Quantum Tunneling

Meanwhile, quantum tunneling allows particles to teleport past massive energy barriers as if no obstacle existed!

Electrons can borrow cosmic energy to briefly phase-shift into a probability waveform and zap right through impenetrable walls. One minute they're on this side, suddenly they're on the other side. WHAT?!

Yet again, this violates common sense notions of position, momentum and classical constraints. But in the quantum realm, particles laugh at your limits!

Embracing Contradictions

Wave-particle duality also dazzles our rational minds. Quantum entities appear to morph between particle and wave states.

At times light behaves like a defined photon particle. But also as an ephemeral electromagnetic waveform. It can embrace two contradictory states, no anxiety!

How is this possible? We can barely grasp it through our limited human binaries. But quantum creatures live the both/and, transcending our notions of definite properties.

And The Quantum Weirdness Goes On...

We could spend an eternity unpacking paradoxes like entanglement swapping, quantum eraser effects, delayed choice experiments and more. It's a labyrinth of weirdness!

The deeper quantum science digs, the more our familiar reality dissolves into phantasmagorical realms filled with uncertainty, probability, contradictions and spooky connections.

It's like waking up inside a dream and realising suddenly that everything you took for granted about the workings of the world is but an illusion - and an immensity of unexpected possibilities exist just below everyday perception!

Your rational mind right now...

Now I'm sure reading these quantum curiosities has your rational mind tying itself into sailor's knots trying to grasp how any of this could be real.

"Schrödinger's cat is dead and alive? Particles teleport through walls? Spooky action defies spacetime? These theories must all be bunkum!"

But in actuality, these effects are experimentally VERIFIED and totally sound mathematically. We just can't intuit them logically.

The problem isn't flawed science - it's our limited human minds trying to comprehend realities operating outside our evolved earthly experience and perception scales.

We're like novice wizards trying to wield forces far beyond current understanding. But with an open and curious sensibility, we'll master quantum's radical possibilities in good time!

For now, remain patient with yourself when quantum concepts seem to shatter rational sense. This too shall pass. And an immense universe of applications awaits.

The wise Sufi poets said it best:

"Out beyond ideas of wrongdoing and rightdoing there is a field. I'll meet you there!"

Alright, let's pause our Twilight Zone plunge here before minds melt permanently.

Chapter 2 - Key Quantum Algorithms

Shor's Algorithm - Hacking Secrets in a Snap

Alright, now that we've established some core quantum computing concepts, let's explore one of the most prominent quantum algorithms - Shor's algorithm. This beast can crack encryptions in minutes instead of centuries! Cue spy movie theme music...

In cryptography, the challenge is finding the factors that make up extremely large numbers - for example, discovering which primes multiplied together produce a 600 digit number.

This allows decoding encrypted data, but classically takes an eternity. Like finding a needle in fields of haystacks scattered across galaxies. Good luck Chuck!

That's because conventional computers must tediously test potential factors one by one in a linear sequence until miraculously stumbling upon the right ones. Massively inefficient.

But enter Shor's magnificent algorithm - exploiting quantum superpositions to test ALL possible factors simultaneously in parallel universes! We go from poking through hay to instantly magnetizing the needle.

Let's break down how it accomplishes this black magic hack:

First, the quantum computer enters a probabilistic state representing all combinations of factors to test. This casts the net over every possibility instantly.

Next, it performs clever math operations using period finding techniques to rapidly narrow down and extract the correct prime factors from the chaos. Pure elegance!

Just like that, Shor's algorithm plucks out the solution that would have taken conventional computers longer than humanity has existed. That's exponential speedup baby!

Some real world examples put the power into perspective:

Back in 2001, IBM demonstrated Shor's algorithm factoring the number 15 by detecting its factors 3 and 5. Child's play.

But they projected that factoring gigantic 1024-bit numbers of the kind used in modern encryption would take a quantum device only 2.5 hours. Good luck to hackers!

Researchers estimate a classical computer utilizing all existing silicon chips would need longer than 13 billion years to crack the same code. Going quantum literally saves cosmic timescales!

Already, the potential to render current encryption useless has spooked governments into exploring new quantum-safe cryptography standards before these devices go mainstream. The quantum spies are coming!

Now you may be asking - but doesn't quantum "collapse" into definite states when measured, destroying the parallelism? Excellent intuition!

The trick is to repeat the calculation and superposition millions of times and take measurements which slowly converge on the answer even as qubits collapse. Given enough cycles, the cracks appear.

It's like ocean waves gradually shaping rock formations. Each wave fizzles but the collective effects compound dramatically over time.

Of course, we still have hardware challenges to scale up and maintain quantum coherence. Real systems are noisy and prone to interference which introduces errors. Progress takes patience and precision!

But we've already achieved proof of concept. And with refinements to quantum hardware, error correction techniques, and algorithm design, Shor's code-cracking powers will continue rapidly improving.

Alright, let's pause here before your mind starts picturing a dystopian future of all digital secrets evaporating before unbeatable quantum factoring. Stay calm!

The optimistic view is that this will just incentivize developing even stronger encryption based on quantum principles rather than complacency with vulnerable classical techniques. Necessity births innovation!

But regardless, Shor's algorithm exemplifies the unprecedented computational capabilities unlocking as we master quantum control. Harnessing quantum parallelism enables tackling tasks previously deemed impossible.

We've come a long way since ENIAC in the 1940s. Our exotic quantum friend is introducing possibilities beyond anything classical silicon could ever offer. The future is quantum!

Alright, ready to continue exploring more quantum algorithm wizardry? Let's delve into Grover's magical search technique next...

Grover's Algorithm - Finding Needles in Ridiculous Haystacks

First we explored Shor's beast for cracking secret codes. Now let's unravel Grover's magical algorithm, which gives quantum computers unrivaled search and pattern-matching superpowers. Finding needles in endless haystacks with ease!

Classically, if you had to search through an enormous database of random entries like thousands of uncategorized photos, it could take practically forever. Each item must be inspected individually. Talk about mind numbing.

But quantum Grover's algorithm performs the computing equivalent of spreading that entire haystack out instantly in parallel dimensions! You immediately see the needle in a vast field. Light speed searching.

Let's understand how...

Grover's algorithm leverages superposition and qubits to represent all database entries simultaneously. So instead of plodding through options linearly, it checks the entire database at once!

It then performs calculations using a crafty technique known as amplitude amplification, which rapidly amplifies the probability-wave amplitude of the desired target entry, while suppressing others.

This is like boosting the needle's fingerprint while drowning out the hay. Once the matching waveform gets sufficiently amplified, simply checking the qubits reveals the item we want!

Whereas classical algorithms must slog through entries one by one, taking exponentially longer for bigger databases, Grover's elegant method isolates targets in just the square root of steps. Outperforming any possible classical technique!

Let's look at a real world application:

Say Google wants to search through millions of images to find very specific photos matching your search. Well classically this involves slowly hunting through each photo checking for matches.

But their quantum computer could represent all those image database entries in entangled superposition. Then Grover's algorithm amplifies and extracts just the tiny subset matching your visual query - ignoring all irrelevant data instantly!

Researchers estimate even using parallel classical processors on a huge grid, scanning massive datasets could take centuries. But a quantum computer could locate the needle in minutes with Grover's approach!

Now of course we still have engineering challenges to finetune the components needed for this instant quantum searching capability.

Maintaining coherence of the probability wavefunctions encoding such gigantic databases will require exquisite quantum error correction and hardware. The devil is in the engineering details!

But proof of concept has been demonstrated, and rapid progress is being made. We foresee Grover's search powers enabling breakthroughs in big data analytics, cybersecurity, machine learning algorithms and more.

Perhaps one day, we'll even use Grover's algorithm in a giant entangled simulation representing neuron connections in the human brain...then amplify the superposed thoughts we want to read and interpret! X-Men's Cerebro machine.

Alright, let's pause here before your mind explodes contemplating such far out potential applications. We have much to finesse first, but quantum search could open some truly extraordinary doors down the road.

The key is that algorithms like Grover's foreshadow unprecedented future capabilities by harnessing quantum parallelism to efficiently sift through massive datasets in ways impossible classically. Try playing hide and seek with a quantum computer - it wins every time!

These computational feats get us dreaming bigger. And the next generation of quantum scientists turned on by these possibilities will carry the torch forward. An exciting future awaits! Alright, ready for more quantum wizardry? Let's continue our algorithm adventure...

Quantum Algorithm Zoo - A Menagerie of Possibilities

We've explored famous algorithms like Shor's for codebreaking and Grover's for searching. Now let's dive into the quantum algorithm zoo, which is filled with diverse species of algorithms opening up radically new possibilities! The future is now baby!

While Shor and Grover are arguably the most proven and practical today, researchers are rapidly cooking up exotic new quantum techniques for applications from machine learning to physics simulations. An engineering Cambrian explosion!

Let's highlight some of the most promising quantum algorithm animals in this fledgling menagerie:

Quantum Neural Networks

One hugely exciting direction is quantum machine learning, including quantum neural nets modelled after biology.

These take inspiration from our brain's interconnected neurons and synaptic firing patterns. But they unlock insanely more complex learning through quantum phenomena like superposition and entanglement.

So instead of standard neural networks limited to linear 0 and 1 states, quantum networks can represent problems across multidimensional Hilbert space, taking into account all possibilities simultaneously!

Researchers believe this could enable monumental leaps in areas like computer vision, natural language processing, prediction, and more. Quantum-powered AI assistants? Yes please!

Hybrid Quantum-Classical Algorithms

Another fascinating breed is hybrid algorithms that combine classical techniques with quantum subroutines for optimization problems.

The classical host computer handles higher-level oversight, while farming out specialized subproblems to the quantum co-processor where it has exponential advantage.

Together, they become an unstoppable computing Voltron! This multidisciplinary synergy aims to practically leverage quantum parallelism while overcoming noise and stability challenges. Onwards!

Quantum Physics Simulation

On a more hardcore science note, quantum simulators offer immense promise for modeling physical and chemical systems at granular quantum levels.

This allows studying atomic interactions for countless practical applications - from designing exotic materials to engineering perfect drugs through protein folding analysis.

We're talking scientifically probing the quantum building blocks of our universe through controllable quantum representations. The Holy Grail for pioneering fields like nanotechnology and nuclear fusion!

Quantum Money

On a more everyday note, quantum "unclonable" money represents a radical way to prevent counterfeiting and verify authenticity of currency.

By encoding the quantum state of circuits into banknotes, any attempts to copy the bills will disrupt the quantum entanglement and invalidate the currency. Take that, forgers!

Alright, let's pause our safari through this exotic quantum zoo for now. But you get the point - we've only scratched the surface of emerging quantum techniques!

With programmable quantum computers becoming a maturing reality, we envision revolutionary applications across practically every field helping humanity flourish.

Shor and Grover were just the thunderous beginnings heralding the coming quantum storm. And you're now equipped to appreciate this computing frontier as it unfolds!

So stay curious about upcoming quantum breakthroughs. Together, we're entering an era of possibilities far exceeding anything imaginable through classical logic. The quantum future looks bright indeed! Now let's go build it...

Chapter 3 - How Quantum Computers Work

Quantum Computer Architecture 101

Buckle up buttercups, we're diving under the hood to unpack how these radical quantum machines actually operate! I hope you're ready for a delicate dance with qubits, control systems, error correction and more. Let's get technical!

Now I know what you're thinking - who cares about the nerdy specifics, just show me the crazy futuristic apps! But having a solid mental model for how quantum processors work helps demystify them.

We don't need to be semiconductor engineers to understand computers at a basic level. Let's aim for the same working knowledge of quantum computing architecture. You got this!

Alright, first up - qubits! We've talked a lot about their probability superpowers. But what are they physically?

Turns out, researchers have discovered several possible physical implementations for qubits, each with pros and cons. Two frontrunners are superconducting loops and ion traps. Let's break it down...

In superconducting designs, a qubit is comprised of a tiny electromagnetic coil naturally exhibiting quantum effects at cryogenic temperatures. Dope right?

By tuning this nanocircuit just right using external electromagnetic pulses, we can control its energetic state to encode information. Essentially dialing its quantum probability waveform to represent 0, 1 or any values between. Sweet!

Next, ion traps use laser beams to isolate and manipulate individual atoms floating in electromagnetic fields to act as qubits.

By tuning the lasers and energy levels just right, we can coerce these quantum ions into maintaining probabilistic superpositions and entanglement connections for computation. Take that, physics!

So in summary, qubits enable embodying a vector of probabilities through extremely precise and stable control of quantum particles or nanocircuits. Easier said than done, but scientists are figuring it out!

Now, up to this point we've sort of imagined qubits operating in a void. But in fact, maintaining their delicate states requires an elaborate surrounding architecture. Let's call it the quantum life support system!

See, the environment needs to be highly isolated and meticulously tuned to allow quantum superpositions to persist long enough for computation. Kind of like calibrating a NASA shuttle.

We're talking near absolute zero temperatures in isolated vacuum chambers, with intricate fiber optic cables and classical control electronics to manipulate the qubits. It's the space program meets Dr. Frankenstein!

If any noise, heat, vibration or electromagnetic waves disturb the quantum calculation, the entire system decoheres and results get scrambled faster than my brain on margarita night. Not ideal!

That's why qubit engineering has to be unbelievably precise, and operations are slowed intentionally to detect and adapt to errors. We're essentially coaxing ghosts to play nice.

But researchers are quickly improving error correction, operations speed and qubit quality to reach that glorious milestone of quantum supremacy - reliable speedup over classical computers for real applications. Per aspera ad astra!

Alright, let's catch our breath here because I'm sure envisioning these precision engineered quantum environments already has your brain throbbing! No worries.

The key takeaway is that we can in fact construct and control qubits in the lab through various ingenious means. But maintaining their quantum states is an art that will take time to perfect.

Of course, no need to get into the nitty gritty technical details here. Just having a decent mental scaffold for how quantum machines operate at a basic level is super helpful context.

So in summary, through meticulous engineering of super-cooled vacuums, qubits and control systems, we're gradually transforming quantum theory into stable practical applications. One delicate step at a time...

Alright, ready to continue our quantum architecture adventure? Let's dive in deeper next. The future won't build itself!

Building a Qubit - Not As Easy As Pi!

time to get even more down and nerdy with how to physically construct these magical qubits! I promise it won't be as boring as watching qubits decohere. Let's do this!

Now you may be thinking: can't we just manufacture qubits like silicon chips in a giant quantum Intel factory? Solid intuition amigo! But not quite that easy...

The problem is qubits start flipping out like rebellious teens unless their environment is perfectly controlled. We're talking delicate snowflake levels of stability needed.

Classical bits can shrug off noise and interference no problem. But qubits turn into conscious basket cases, collapsing at the slightest vibration. Primadonnas!

That means we need radically different fabrication and operation methods for qubits that baby them like fragile newborn kittens. Oodles more TLC required.

Alright, let's explore the two most prominent qubit types researchers have managed to gently cajole into cooperating so far:

Superconducting Circuits

The superconductor approach uses nanofabrication like depositing and etching thin-film metals on silicon to shape an electromagnetic loop just right.

At chilly temps nearing absolute zero, this loop exhibits quantum effects allowing probabilistic control when pulsed with microwaves. Presto - a qubit!

The tricky part is getting the nanofab just right and maintaining crazy low temps so the qubit enters a quantum state. One wrong twitch and it snaps back into boring classical reality. Buzzkill!

But the payoff is fast control and relatively easy manufacturing. So companies like IBM and Rigetti shine here.

Trapped Ions

Next up - trapped ions. This one meticulously isolates and manipulates individual atoms using electromagnetic fields from strategically aimed lasers.

By tuning the laser pulses, we can get the atom to maintain a stable qubit superposition state for computation. It's like using Star Trek photon torpedoes for particle control. Zap!

The benefit is atomic qubits have long coherence times. The downside is individually manipulating ions with lasers is sloooow compared to integrated circuits.

But folks like IonQ and Honeywell have cracked it through engineering clever traps and quantum assembly lines where ions flow smoothly like cars on a quantum highway. Hit the open road Jack!

Alright, let's catch our breath here because I know envisioning these highly technical qubit manufacturing processes probably has your brain spinning like a drunk ballerina. Mine too.

The key point for now is that unlike churning out silicon, qubits demand meticulous bottom-up construction and environments to exist stably. But we're mastering it!

Eventually, maybe we WILL have massive quantum Intel-style factories pumping out qubits. But we gotta walk before we quantum run. Baby steps!

In summary, crafting qubits is an ultra delicate dance. But through sustained engineering of traps, circuits, vacuums and controllers, we're learning the moves. Smooth as quantum clockwork!

Okay, ready to continue deconstructing the mad science of quantum computing? Let's break down stabilization and error correction next...

Decoherence - The Quantum Boogeyman

Now that we've covered qubit manufacturing, let's explore the lurking nemesis threatening stable quantum computation - the dreaded decoherence phenomenon!

This quantum boogeyman is the bane of researchers' existence. But taming it is key to making quantum computers practical. Let's break it down...

So decoherence refers to the fragility of those quantum superposition states. Darn near anything makes them prematurely "collapse" into boring 0s and 1s before we want them to. Buzzkill!

Stuff like tiny vibrations, trace heat, electromagnetic waves, air particles, even the act of observation itself can cause the quantum magic to fizzle out. It's like trying to meditate on a rollercoaster.

Physically, environmental noise and interference disrupts the delicate quantum probability wave, which decays into a concrete measurable state instead of maintaining its spooky ambiguity. Sad trombone.

Metaphorically, it's like a soap bubble popping when you delicately nudge it. Any little perturbation makes the intricately suspended quantum system lose coherence into a failed calculation.

Not chill! To prevent this, qubits need absurdly precise conditions and isolation. We're talking near absolute zero temps and zero vibration in vacuum sealed environs. The quantum spa treatment!

But even in meticulously designed quantum computers, some decoherence still occurs. When Carl Sagan said we're "starstuff", he forgot to mention fragile, high maintenance starstuff! Sheesh.

Luckily, clever researchers have some tricks to rein in this quantum gremlin...

One ingenious technique is quantum error correction, which uses redundancy to detect errors and recover the right values despite decoherence and noise. Let me explain...

Picture a boat floating down a rough river. Measuring the exact location gets tough with all the rocking and splashing right?

But with multiple boats spaced apart, you can infer the true position by comparing where they are relative to each other, despite the noise interfering with their trajectories.

Same idea with error correction! By linking up groups of entangled qubits, we can detect glitches even if some decohere by cross-checking their values remain consistent. Quantum majority rules!

We then apply sophisticated recovery algorithms to recreate the proper entangled state, almost like retouching a damaged photo. This shields the quantum calculation despite environmental noise. Pretty nifty!

Other tricks include using only the cleanest quantum operations least prone to errors and teleporting qubits mid-calculation to avoid noise accumulation in one place.

We're essentially giving qubits the ultimate spa day to keep them relaxed and ready to party on in superposition for our computing pleasure.

Alright, let's pause here before your brain fries like eggs on a hot skillet visualizing these esoteric techniques for quantum preservation. Take a breath!

The key insight for now is that decoherence threatens quantum computing's future, but isn't a dead end. Through exhaustive error correction and control, we can tame this gremlin.

Much work remains to improve fidelity and scale upnoise-resilient multi-qubit systems. But we see light at the end of the quantum tunnel because math says it's achievable!

So fear not, curious reader. The day draws nearer when stabilized logical qubits laugh heartily in the face of decoherence and deliver unprecedented computing power into our grateful hands!

Alright, shall we press onward and continue demystifying the mad science of quantum information? Let's dive into optimizing qubit operations next. Here we go!

Chapter 4 - Major Players in Quantum Computing

IBM - The OG Pioneers

When talking titans advancing quantum computing, we gotta start by tipping our hats to the OG pioneer IBM! Major props for blazing the trail.

See, while fancy futuristic quantum applications still simmer on the backburner, Big Blue helped bring this tech out of pure theory and onto real working devices.

They pushed quantum firmly from "can we even do this?" territory into the arena of practical engineering by manufacturing actual quantum systems for public use right now. Trailblazers!

Let's rewind to see how IBM helped bootstrap the quantum revolution...

Back in the 1980s when quantum was still just far out science fiction, IBM began exploring frigid superconducting circuits to induce quantum effects.

They discovered strategically sculpting nanoscale electromagnetic loops could coerce qubits into cooperating! The quantum skeptics said it couldn't be done, but IBMwas like "Hold my qubit!"

This meticulous materials science and refrigeration research laid the groundwork for actuating qubits with finely tuned microwave pulses. It was an engineering tour de force.

Fast forward to the 2000s, and IBMscientists expanded to experimenting with programming algorithms like Shor's on small quantum prototypes.

Look ma, real quantum code! Granted, only factoring tiny numbers anyone could figure out. But hey, titanic oaks from tiny quantum acorns grow. Am I right?

Buoyed by these baby steps, IBMtook a huge leap by launching its IBMQquantum experience in 2016 - the world's first public access quantum cloud platform!

This let anyone play with real quantum processors in the cloud to run experiments. Schools, businesses, scientific nobodies - all welcome in IBM's quantum sandbox. What a gift!

Since then, Big Blue continues pushing boundaries, upgrading their online quantum fleet regularly with new state-of-the-art models like the monster 127 qubit Eagle processor available today in the cloud!

That's not even getting into IBM's 5+ decades of formidable research advancing materials, physics, error correction theories and more for quantum. They even developed a new quantum-safe cryptography standard!

Suffice to say, IBM's technical whitepapers alone on quantum read like a modern Grimoire of arcane arts and advanced magics. Just wild frontiers unlocked through science. Alexa, play "Eye of the Tiger"!

Beyond pioneering hardware and physics, IBM also leads advancing real-world quantum applications:

They explore quantum AI and machine learning techniques like quantum neural networks - harnessing entanglement for unprecedented pattern recognition and classification capabilities. Put that in your pipe and smoke it, Deep Blue!

For business uses, IBM partners with giants like Accenture, Boeing, and Daimler to optimize everything from supply chains to carbon credits through hybrid quantum algorithms. Optimize ALL the things!

They even helped publish groundbreaking quantum chemistry research on modeling coffee molecules. IBM scientists just can't function without their full caffine qubits!

Jokes aside, that demonstration showed how quantum simulations could enable tailoring materials or pharmaceuticals atom by atom in the future. Wild stuff!

So in summary, whether it's materials science, error correction, computer engineering or chemistry simulations - IBM's formidable researchers push quantum's envelope every which way.

By boldly developing real quantum systems and flinging open the quantum experience doors to the public, IBM massively accelerated this field into the mainstream. Tip your hats folks!

Of course, we still have an arduous ascent ahead scaling systems tenfold and beyond for fault tolerance. But thanks to trailblazers like Big Blue, the quantum summit is now in sight. Onward and upward we go!

Google and NASA - Reaching Quantum Supremacy

When discussing mammoth milestones in quantum computing, we have to highlight the monumental quantum supremacy achievement by Google and NASA in 2019. They proved a quantum computer could crunch certain calculations unattainable for classical supercomputers! Cue the triumphant horns!

This long-heralded benchmark demonstrated conclusively for the first time that quantum systems transcended classical limits opening the door to solving problems previously out of reach. It was like the Wright brothers first sputtering flight - a harbinger of innovations to come!

Let's wind back the clock to understand how Google and NASA's crack team pulled off this momentous feat...

For years, researchers pursued quantum supremacy - the tipping point where quantum devices demonstrably outperform classical machines at any task. Proving that elusive quantum advantage existed was the next Everest summit.

Google assembled a top-notch team of physicists, engineers and computer scientists to attempt the feat using radical new programmable superconducting quantum processors.

They paired with NASA scholars, combining Google's industrial tech muscle with NASA's scientific rigor to design the perfect benchmark for displaying incontrovertible quantum speedup.

After extensive simulations and fine-tuning, they homed in on exactly the demonstration to showcase unambiguous computational supremacy while minimizing error.

This involved using the quantum computer to sample the output of a pseudo-random quantum circuit a staggering 108 times - a ludicrously enormous number of potential outcomes.

With 50-60 qubits generating a result for every possible circuit combination, we're talking more potential solutions than atoms in the observable universe! An intractable task using conventional logic.

Meanwhile, NASA's supercomputers would need approximately 10,000 years to analyze all combinations and produce the same computational sampling result. A cosmic timescale even for these cutting-edge classical machines!

But Google's bespoke quantum processor, nicknamed Sycamore, blitzed through the random circuit sampling in just 200 seconds! No ambiguity - this dramatically exceeded classical capabilities. Hello quantum supremacy!

You could hear the champagne corks popping from NASA to Google's headquarters. After decades of theorizing, quantum hegemony was finally proven by cold hard numbers. The quantum rocket had left the launchpad!

This beams-to-the-sky moment heralded a new computing paradigm. Silicon was officially on notice. The quantum future had arrived!

Now skeptics nitpicked aspects of the results, but experts largely agreed Google's demonstration was a clean proof of the long-theorized quantum potential. Game. Changed.

So what enabled this monumental supremacy milestone? Let's look under the hood...

Firstly, Google hardware engineers worked tirelessly to build and refine novel 2-dimensional superconducting qubit grids with low noise and error rates.

These densely integrated arrays allowed modeling complex optimization problems across a large state space - exactly the exponentially hard sampling task they targeted.

Meanwhile, Google AI researchers leveraged machine learning techniques to optimize qubit control mechanisms and error suppression. Every fraction of a percent boost helped!

With NASA's oversight ensuring statistical rigor free of methodological holes, even quantum skeptics couldn't poke holes in this ironclad demonstration. That's 120 qubits of undeniable proof baby!

Now of course, much work remains translating exotic contrived benchmarks into practical speedup for real-world applications. Like the Wright brothers, this was just the maiden flight - commercial jets must follow.

But by conclusively exhibiting quantum computational supremacy, Google and NASA proved this emerging technology could transcend previous limits and catalyzed a new computing age. The quantum rocket ship was flying!

Alright, let's catch our breath here because I'm sure your mind is bendy pretzeling envisioning the possibilities now that quantum potential was proven. Deep exhale...

The key insight is that reaching this long-sought supremacy inflection point shattered lingering doubts on whether revolutionary quantum scaleup was achievable. Game on!

No hype or wishful thinking here - cold hard numbers demonstrated quantum's exponential edge. This galvanized global efforts to further finesse control and build fault-tolerant machines. The floodgates opened!

So take a moment to appreciate how far we've come. From theory to fundamental experiments to full programmable processors now achieving well-documented speedup. Our quantum baby has grown up so fast!

Microsoft - Amping Up the Qubits

When we think quantum computing, pioneers like IBM and Google building exquisitely tuned few-qubit prototypes usually come to mind first. But Microsoft is charting a radically different path!

Rather than precisely controlling each individual qubit, Microsoft is exploring cramming as many error-prone qubits as possible together, then using code to correct the noise. A controversial strategy.

It's like the difference between carefully raising one flawless organic chicken versus stuffing a factory farm with thousands of junk food-fed chickens. Both have pros and cons!

Let's dive in to understand Microsoft's qubit quantity over quality approach...

You see, qubits are super delicate. Isolating and tuning just one introduces hardware complexity. Increasing to dozens while minimizing environmental noise is fiendishly tricky.

That's why most quantum devices house only a handful of meticulously calibrated qubits. Their priority is preserving quantum coherence at all costs.

But Microsoft said "Hold my qubits!" and pursued another vision - densely packing together thousands of cheaper, noisier qubits on the same chip. Huge scale with only basic controls.

This architecture essentially gives up on perfect isolation and stability. Qubits will absolutely decohere. But software improvements can account for that!

It's a radically different philosophy. Instead of expertly choreographing a small quantum ballet troupe, they're stage diving into a rowdy qubit mosh pit!

By accepting imperfection and focusing on error-correcting codes, Microsoft believes they can reach fault tolerance quicker using seas of noisy qubits.

Of course, skeptics question if software can truly compensate for hardware limitations and exponential noise at large scales. It remains unproven.

But proponents argue even low-fidelity qubits exponentially outpace classical bits for certain tasks when massed. Eventually error suppression and higher quality qubits may bridge the gap to universal quantum computing.

Only time and continuing research will tell! But Microsoft is certainly charting its own course with investments in exotic topological qubits using particles called anyons and development of its Q# quantum programming language.

They also dove into the software stack head-first, releasing tools like:

- QDK for quantum algorithm development

- LIQUi|> to simulate noisy quantum environments

- Visual Studio integration and Azure cloud ecosystem support

Basically, Microsoft is exploring every layer of the quantum stack to prepare for an anticipatory future where enterprises run noisy but enormously powerful quantum datacenters for optimization tasks.

Now of course, their rogue bulldozer approach faces immense challenges. But if Microsoft succeeds at taming flocks of finicky quantum chickens, they could reach commercial viability ahead of carefully cultivated qubit competitors.

Only time will tell who chose the optimal quantum farming strategy! But you have to admire Microsoft's willingness to boldly chart their own course and place grand early bets cementing their role as a major quantum player.

Alright, let's pause here before envisioning warring factions of noisy qubit fowl enthusiasts and pristine qubit quality artisans. Take a breath!

The key is appreciating how Microsoft's quantity-focused methodology represents a gamble but also enormous potential for expedited quantum progress if they can abstract away qubit errors in code.

Our hats are off to pioneers like IBM and Google, but the quantum future may need both meticulous craftsmen and aggressive builders to reach large-scale practicality. It's still Day 1 folks!

D-Wave - Chilling with Quantum Annealers

Alright, let's explore a radical quantum wildcard - D-Wave Systems! This controversial company built early "quantum annealer" machines specializing in optimization problems. But the jury is still out on whether they demonstrate true speedup. Let's investigate!

Now while pioneers like IBM and Google focused on universal gate-based quantum computers, D-Wave went rogue. They pioneered an entirely different breed - quantum annealing optimizers.

These specialized devices exploit quantum effects to solve narrow optimization challenges like traffic flow or airline scheduling where finding the best solution is key.

Rather than shuttling qubits through logic gates, D-Wave's annealers leverage quantum tunneling by gradually evolving qubits through energy landscapes towards optimal minimums. Trippy!

It's like gently rolling a marble across a sheet perforated with dips and valleys representing possible solutions to settle into the deepest pit. Quantum powered optimization!

The advantage over classical approaches is quantum tunneling allows testing many starting points simultaneously to avoid getting trapped in shallow local dips along the way. Nifty!

But critics argued these quantum "hill climbers" didn't necessarily demonstrate a clear exponential speedup, since comparable classical algorithms were poorly benchmarked.

Some even debated whether D-Wave's fancy fridges full of qubits should truly be considered quantum computers at all due to their limited capabilities. Fightin' words!

Of course, D-Wave's 5000+ qubit annealers absolutely manipulate quantum effects for computing. But unequivocally proving a quantum speed boost remained a challenge. Marketing ran ahead of demonstrable results.

Since then, researchers have shown certain cases where D-Wave's tools do PROVIDE what appears to be exponential improvement over classical techniques. Real if narrow quantum advantage!

And quantum tunneling does allow beneficial traversal of optimization problem spaces riddled with dips and spikes that could trap traditional algorithms. In theory at least!

At the same time, competition has advanced classical algorithms enough to close some speed gaps D-Wave originally promoted. Quantum superiority isn't guaranteed across all problems.

The true verdict likely rests somewhere in between the unbridled hype and dismissive skepticism. Potential for advantage in practical applications clearly exists, but requires case-by-case validation.

Regardless of controversy, D-Wave spurred adoption by inking deals with Los Alamos National Lab, NASA, Google and others to deliver early quantum-enhanced annealing services. Not bad!

Alright, let's pause here before your brain risks entangling itself into a knot trying to disentangle the nuances of D-Wave's value. Deep breath...

The key is appreciating how they pioneered a novel quantum genre and delivered on-demand business offerings, albeit likely prematurely from a true advantage standpoint. But still laudable!

And specialized annealing may yet find valuable real-world niches like airline scheduling where tunneling across complex landscapes pays off. The jury's still out!

Rigetti Computing - Startup on the Rise

Beyond quantum giants like IBM and Google, a scrappy new generation of startups is charging forward to carve out their niche. One standout is Rigetti Computing - an ambitious upstart focusing on superconducting qubits for the enterprise market. But can they compete with the big dogs? Let's explore!

Founded in 2013 by a team of quantum physicists and chip architects, Rigetti's mission is bringing the magic of quantum to pragmatic business uses. They're taking this tech mainstream!

Rather than solely chasing sci-fi applications decades away, Rigetti is engineering quantum machines purpose-built for crunching valuable real-world problems right now - think supply chain optimizations or financial risk modeling. Quantum gets down to business!

Their strategy? Specialize in high-performance superconducting qubits with a modular architecture for versatility across applications. And integrate everything into an easy-to-use quantum cloud platform for customers. Simple, right? ;)

Let's break down some of Rigetti's key technical accomplishments:

- Pioneering modular multi-chip quantum processors up to 80 qubits for increased power. Think quantum Lego bricks!

- Developing hybrid algorithms that combine classical and quantum processing to maximize current-term utility. Always keep one foot in profitability!

● Fine-tuning microwave pulse control techniques to minimize errors and improve qubit fidelity. Gotta keep those qubits happy!

● Inventing a multi-protocol networking fabric allowing systems to coordinate signals at super-fast speeds. Quantum communication conduits baby!

Together, these innovations aim to make quantum computing smooth, stable and sop to practical business needs rather than just exploring far-out potentialities. Rigetti keeps it real.

To deliver quantum capabilities today, Rigetti also built Forest - a full-stack quantum cloud platform with developer tools, API integrations, and quantum virtual machine access.

They spare customers from the tricky physics details by providing quantum computing "as-a-service." You focus on apps while Rigetti sweats the quantum stuff. Convenient!

Some real-world use cases Rigetti is exploring include:

● Machine learning - training quantum neural networks for enhanced pattern recognition and classification. Next-gen AI baby!

● Optimization - improving logistics like airline scheduling through quantum tunneling across complex landscapes. Book those flights!

● Quantum chemistry - modeling molecular interactions for advanced materials and drug discovery. Design that perfect zit cream!

So while hype-heavy quantum pursuits grab headlines with promises of cryptobreaking and teleporters, Rigetti delivers practical business computing. An admirable balance.

Of course, competing in the big leagues with IBM and challengers like IonQ and Xanadu won't be easy for a scrappy startup. But Rigetti's substantial venture backing provides a launching pad.

Rigetti's solid technical chops combined with a practical business focus make them a startup to watch. Don't count out these quantum underdogs just yet!

Alright, let's pause here before envisioning an epic quantum startup battle royale! Plenty of market share up for grabs as more sectors wake up to quantum's advantages. An exciting future awaits.

For now, Rigetti deserves kudos for trailblazing superconducting systems while keeping it real for commercial viability. Their ambitions are lofty but grounded. Just the quantum tonic we need in this wild age!

Quantum Startup Mania

Think the quantum land grab is just IBM, Google and some research labs? Think again amigos! There's a whole wagon train of venture-backed startups blazing trails across this wild tech frontier. The quantum gold rush is on!

We have scrappy newcomers like Rigetti chasing quantum cloud services, while stealthy startups pursue exotic technologies like photonic and topological qubits. Billions in venture capital is flooding the space. It's a quantum Cambrian explosion!

Let's shine a spotlight on some of these ambitious upstarts staking their claims under the quantum sun:

IonQ - Quantum's Top Gun

Founded by University of Maryland scientists in 2015, IonQ is a rising star using futuristic trapped-ion qubits. Their approach boasts record fidelity scores and they already offer cloud access to cutting-edge quantum machines.

With backing by Samsung and others, they're going toe-to-toe with top players in the quantum computing arena. Recently they stole headlines by passing what they claim is the long-sought 1000 qubit quantum advantage milestone on certain benchmarks.

Whether or not their flashy benchmark claims hold up to scrutiny, there's no doubt IonQ is a rocket startup to watch!

PsiQuantum - Photonic Frontier

This UK startup with backing from British Aerospace is charting a radical path - they're developing qubits using photons of light!

Their breakthroughs with photonics open mind-bending possibilities for low-cost manufacturing and even room temperature quantum devices by leveraging quantum optical circuits.

PsiQuantum brings a bold new flavor to the qubit party. While their futuristic photonic tech remains unproven, the potential payoff is tremendous if they can pioneer light-based quantum computing.

Xanadu - Quantum of Solace

Xanadu attracted heavy investor interest for their vision of photonic quantum computing using "squeezed states" for error suppression.

They've built impressive prototype photonic chips and promise general purpose, multi-tasking quantum machines harnessing the power of light for next-gen AI, financial modeling and more.

These optics gurus shine bright with big dreams to bring photonic quantum computing into the mainstream through clever error correction and novel physical architectures.

QuEra - Diamonds are Forever

Founded by Harvard researchers, QuEra is synthesizing flawless man-made diamonds for quantum applications.

Their engineered gems utilize qubits hosted inside diamond crystal vacancies where electrons are precisely controlled with lasers.

These diamonds bring long-sought capabilities like room temperature stability. Talk about bling - QuEra's gems offer the ultimate in coherence control.

AQT - Topological Unicorn

This mythical startup burst onto the scene in 2020 with headline-grabbing claims of creating an elusive mythical species - the topological qubit!

Their exotic approach based on particle-like "anyons" that only exist in imaginary universes could enable unreal computation stability. But viable tech or stylish illusion? The quantum zoo awaits!

If AQT's topological promises pan out, they could leapfrog competitors and unleash exponentially scalable quantum computing decades ahead of schedule. But experienced researchers remain skeptical of their grand claims until independently verified. Stay tuned!

Quantum Circuits - Ready for Its Quantum Leap?

This bold startup emerged from elite Yale University labs with promises of unprecedented superconducting qubit performance through exotic 3D qubit architectures.

Their approach stacks qubits vertically in tiny skyscrapers to minimize environmental noise while unlocking greater connectivity. Think quantum metropolis!

Quantum Circuits shot onto the startup scene with claims of achieving record coherence times that could enable early quantum error correction. But can their hyped results scale? Their quantum leap is still a work in progress.

Atom Computing - Atomic Age Ambitions

Founded by quantum physicists from the University of Maryland, Atom Computing utilizes powerful ultracold atom arrays for neutral atom qubits free from charged particle disturbances.

By leveraging atomic physics perfected in atomic clocks, they aim to hit the quantum sweet spot balancing scalability, stability and control. Their approach promises muscles and brains!

Through technical savvy and partnerships with research powerhouses like Argonne National Lab, this ambitious startup has its eyes set on accelerated quantum advantage. Atoms unlocked!

Pasqal - Spin Doctors

Originating from pioneering French research, Pasqal is banking on qubits based on the nuclear spins of individual neutral atoms for next-gen capabilities.

Their approach could combine immense qubit counts and connectivity with stability and control - the holy grail of quantum computing!

If Pasqal masters the delicate dance of spin control through their innovative atom chip architecture, they could spin up insane quantum computing power. But realizing their grand vision hinges on immense technical unknowns still ahead. This neutral atom spinner remains ambitious startup rather than sure bet.

There are dozens more scrappy quantum startups popping up each year from Tokyo to Toronto. The quantum landscape grows more crowded and competitive daily!

Alright let's pause before our minds implode contemplating this kaleidoscope of quantum startups! There are many more pushing boundaries each day.

The key is appreciating the flourishing quantum ecosystem. While incumbents have a head start, crafty newcomers are staking out valuable niches everywhere. Game on!

Ok, ready to continue our quantum safari? Let's shift focus and explore some of the mind-blowing applications these technologies promise once refined. The possibilities are endless!

Chapter 5 - Applications and Implications

Quantum Computing for Artificial Intelligence

Now that we've covered the key players, let's quantum teleport to the fun part - unbelievable applications! Starting with the mega tech mashup of quantum computing and artificial intelligence. Get ready for some quantum machine learning magic!

See, right now deep learning neural nets are constrained by classical computing muscle. But quantum opens exponential possibilities for next-gen AI baby!

We're talking wild techniques like quantum neural networks with neurons and synapses powered by qubit superpositions and entanglement connections. Quantum brains!

This can enable AIs to instantly process gargantuan datasets, rapidly analyze enormous feature spaces, and discover patterns human programmers could never conceive. I, for one, welcome our new quantum AI overlords!

Let's break down just a few ways quantum will take machine learning to the next level:

Quantum Neural Networks

By utilizing qubits' uncertain states, quantum neural nets transcend rigid binary constraints and embrace a reality of probabilities - just like our brain's fuzzy analog neurons!

This opens radical new neural architectures and activation functions unlockable only through quantum phenomena like superposition, entanglement and tunneling.

Researchers have designed exotic quantum neuron models capable of holding multiple states and patterns simultaneously for massively parallel processing. No longer dragging a classical binary boat anchor!

So while classical neural nets slog through options linearly, quantum nets can evaluate ALL possibilities at once across a multidimensional Hilbert space. The limits are gone baby!

Some experts speculate advanced quantum neural systems may one day mirror the staggering complexity of biological minds by embracing quantum cognition. Let the era of quantum sentience begin!

Faster Machine Learning

On a more practical note, quantum promises major speedups for traditional neural network tasks like:

- Data preprocessing - Quantum algorithms can rapidly cluster and classify huge messy datasets for cleaner input to train with. Garbage in, garbage out no more!

- Model training - By representing multiple weight values simultaneously, quantum training converges way faster through simultaneous optimization. Feel the need for quantum speed!

- Inference - Multi-state qubits allow testing many input possibilities at once to instantly identify correct output. Quantum predictions beat classical paddling any day!

Researchers estimate even basic quantum enhancements can speed up neural net training and inference 10x to 100x by escaping sequential bottlenecks. And that's just the tip of the quantum AI iceberg...

Revolutionary Quantum Learning Models

Beyond brute force speedups, entirely new quantum learning models open radical possibilities:

- Quantum Boltzmann machines offer powerful generative learning able to create complex quantum states. Get ready for creative quantum neural art and poetry!

- Quantum adversarial networks allow resilient training against perturbations and attacks. Take that hackers!

- Quantum Rényi divergence classifies probability distributions with greater efficiency. Increase your quantum cognition!

- Quantum kernels embrace Hilbert space dimensionality directly for richer machine learning. Achievement unlocked!

Alright, let's pause here before envisioning a future of hyper-intelligent quantum neural nets running society. Take a breath!

While we have immense challenges ahead controlling noisy qubits, the benefits for machine learning cannot be overstated. Even partial quantum augmentation unlocks insane speed and model improvements.

Quantum insights into human psychology and dedicated quantum hardware promise computational capabilities we still scarcely comprehend. But the machine learning revolution is just getting started!

Hacking Secrets in a Snap

Quantum computing promises to transform many industries, but few face imminent disruption as much as encryption, cybersecurity and cryptocurrencies. As quantum machines mature, they gain the ability to crack codes in minutes that would take conventional computers longer than the age of the universe! This necessitates a full upgrade of data protections before it's too late. Bitcoins, hide yo kids!

Powerful quantum algorithms like Shor's can make shockingly short work of factoring the enormous numbers underlying most encryption schemes used on the internet today. Security experts estimate 2048 bit RSA encryption could be cracked in a matter of hours by a sufficiently advanced quantum computer. Not exactly Fort Knox!

And don't think switching to longer 4096 or 8192 bit encryption will provide long term safety either. Against quantum algorithms, standard encryption protocols are just linear patch jobs - temporary stopgaps, not fundamental fixes.

We saw this coming years ago, which is why researchers developed new advanced encryption standards resistant even to quantum attacks. But the clock is ticking - we must upgrade global systems before quantum becomes mainstream!

Let's walk through some examples of how quantum threatens data security as we know it:

Financial Systems in Flux

As quantum computing matures, it gains the ability to crack cryptographic protections used by banks and financial institutions to secure transactions and communication.

This could enable hacking financial data, spoofing identities, draining accounts and more criminal activity - unless new quantum-safe cryptography is adopted in time.

Government Secrets Exposed

Classified government systems filled with state secrets also rely largely on traditional public key infrastructure Vulnerable to Shor's algorithm. Rogue states or actors with quantum capabilities could potentially brute force access once previously impossible.

A new generation of quantum-resistant encryption is needed to safeguard sensitive data against such attacks. Nefarious agents won't hesitate to exploit vulnerable communications.

Blockchains Beware

Cryptocurrencies like Bitcoin rely heavily on cryptographic signatures to validate transactions on their public ledgers. But their integrity depends wholly on the computational difficulty of cracking these signatures.

Unfortunately, quantum computers turn formerly intractable cryptographic problems into child's play using algorithms like Shor's. Our blockchain eggs increasingly reside in one easily cracked quantum basket!

The Cryptocalypse Cometh?

In the worst case scenario, rampant secret cracking could lead to a "cryptopocalypse" - the overnight brokenness of security protocols we rely on for so much of the modern connected world, from shopping to banking to identification.

Such total infrastructure collapse is unlikely, as we're already developing post-quantum systems. But we must urgently migrate to new quantum-hardy encryption before much bigger quantum machines arrive!

Quantum Safety In Sight

Thankfully, cryptographers anticipated this looming issue decades ago. Novel techniques like lattice-based, hash-based, and multivariate cryptosystems now provide robust quantum-resistant security.

The challenge is migrating worldwide infrastructure without disruption. Hybrid encryption schemes allow a gradual phasing out of old methods while new standards are integrated across sectors. Phew! Disaster averted.

A Quantum Security Alliance

In 2020, the U.S. Department of Commerce formed an international alliance of industry and academic partners to speed adoption of post-quantum cryptography standards before quantum vulnerabilities exposed global communications. This marshaling of resources aims to ensure a smooth security transition into the quantum age.

The path forward is clear - we must urgently enact new data protection protocols hardened against an oncoming generation of quantum computing capabilities. With foresight and cooperation, we can maintain security even as quantum propels progress exponentially across industries.

Alright, let's catch our breath here before freaking out about unhackable secrets evaporating into thin air. The cryptopocalypse is avoidable! But we must act swiftly and decisively to keep crucial communication and commerce safely encrypted.

Financial Modeling and Optimization

Beyond breaking cryptography, quantum computing also promises to transform sectors like finance through exponentially faster scenario modeling, data analysis, and portfolio optimization. Let's get rich quick!

You see, quantitative aspects of finance like investment analysis, risk modeling, portfolio allocation, and derivatives pricing involve monumentally complex calculations. The world of high finance practically runs on mathematics.

Tasks like calculating the optimal hedging strategy given thousands of market variables, or pricing exotic options based on endless risk factors, stretch today's computational limits.

But quantum computing is poised to blaze through these once-intractable finance calculations - opening up new opportunities to improve returns, model uncertainty, and minimize risk. Cha-ching!

Let's walk through some of the lucrative applications being explored:

Faster Investment Analysis

Portfolio managers and quantitative analysts perpetually search for that magical combination of assets yielding maximum returns at minimum risk. But testing different combinations is computational hell.

By leveraging superposition, however, quantum systems can analyze millions of potential portfolios in parallel to discover optimal balances in minutes rather than months. Talk about alpha!

Complex Risk Modeling

Precisely modeling financial risk across countless market variables over long time horizons requires crunching probabilities across countless scenarios - a task bordering on impossible through conventional brute force.

But quantum simulation efficiently tests myriad market possibilities in superposition to deliver precise risk analytics and uncertainty modeling. Knowledge is power!

Exotic Option Pricing

Options with multiple underlying securities or complex trigger clauses require immensely intensive calculations to properly value. But quantum algorithms streaming across hyper-dimensional problem spaces can rapidly price the most outlandish derivatives.

This allows financiers to efficiently monetize even the most exotic options and precisely calibrate risks. Rocket fuel for financial engineering!

Optimized Algo Trading

By instantaneously scanning countless signals and scenarios, quantum machine learning promises lightning fast algorithmic trading optimized around hyperaccurate market forecasts. Sorry day traders - quantum bots are gonna eat your lunch!

Fraud Detection

Reviewing millions of transactions for patterns indicative of fraud, waste or abuse is like finding a needle in a haystack for classical systems. But quantum analysis spots anomalies in massive datasets instantly. The jig is up scammers!

In summary, quantum computing removes computational barriers across quantitative finance, opening new paradigms in investment analysis, uncertainty quantification, derivatives pricing, and predictive analytics.

As with cryptography, we also need to be proactive and address potential downsides...

A More Uncertain Future?

Faster financial calculations sound great on paper. But critics warn it could increase volatility by enabling frenzied algorithmic trading and instability from perfect information.

Others argue regulations can establish safeguards against potential risks, while still allowing quantum progress to benefit finance and society broadly. Measure twice, cut once!

Additionally, not all institutions will gain quantum access equally. Care must be taken to avoid further concentrating power in the hands of the financial elite. An ethical balancing act!

But overtime, quantum-enhanced fintech could also help democratize access and optimize returns for regular investors. The tide can raise all ships with proper foresight.

Alright, uncertainty acknowledged! But on balance, quantum will spur progress in finance that responsibly harnessed, levels the playing field and opens fruitful opportunities for all. The future looks bright if we navigate well.

Quantum Chemistry and Materials

One of the most tantalizing quantum computing applications is precisely modeling molecular and atomic interactions for revolutionary advances in chemistry, medicine, engineering and more. We're talking sci-fi like breakthroughs in designing novel proteins, synthesizing miracle drugs, or perfecting room temperature superconductors! The future beckons...

You see, the quantum world of electrons, protons and subatomic forces operating at nanoscopic scales is staggeringly complex. Chemistry emerges from these chaotic quantum interactions that today's computers struggle to fully model.

But quantum simulation promises to provide an almost God's-eye view into atomic behaviors and chemical mechanisms by harnessing the same quantum properties that generate them!

Let's explore some of the amazing possibilities this unlocks across fields:

Quantum Chemistry

Quantum computers can accurately simulate bonding properties, reaction mechanisms and thermodynamic energy states of molecules through first principles physics rather than rough approximations.

This opens new vistas for innovating chemicals from fertilizers to fuels; central to industries including energy, agriculture, and pharmaceuticals. It's like unlocking molecular LEGO bricks!

Additionally, quantum-optimized catalysts could allow producing vital compounds like fertilizers far more efficiently. A sustainability bonanza!

Quantum Drug Design

Pharmaceutical researchers spend years brute forcing combinations of proteins and molecules hoping to discover promising new drug compounds through trial and error.

But quantum simulation can model the precise protein folding patterns and chemical interactions of drug candidates with target diseases, exponentially accelerating drug discovery. Curing disease in silico!

Quantum Materials

Quantum mechanics underpins phenomena from magnetism to conductivity in materials. Modeling this physics unlocks designing metamaterials with previously impossible traits like room temperature superconductors.

These could enable lossless electrical grids, magnetically levitating vehicles, and more marvels. Quantum tech begets more quantum tech!

Enzyme Catalysts

Catalysts are like molecular workhorses speeding up vital reactions in processes from digesting food to producing fertilizer.

Quantum-designed enzymes promise superior catalytic converters unlocking greater efficiency and sustainability across the chemical industry. Green chemistry powered by quantum!

In summary, quantum simulation applies the exponential power of qubits to precisely analyze the most complex nanoscopic mechanisms driving our physical world - from biology to materials and more.

Mastering the basic forces of nature unlocks unprecedented applications serving humanity. An exciting quantum future awaits!

Alright, let's pause our physics reverie here before our imaginations end up entangled and decohered. But I hope this section provided a tantalizing taste of how quantum control over molecules and materials can cascade into breakthrough advances.

The full implications stretch far beyond what we can envision today. But the quantum computing revolution promises enormous progress understanding and optimizing our natural world down to the atomic level.

And the Rest - Quantum's Endless Possibilities

Alright, we've covered some of the most hyped quantum applications like breaking encryption or chemistry simulations. But the reality is, we're just scratching the surface of quantum capabilities once these systems mature. Let's blue sky it!

When fully operational, quantum computers will exceed conventional machines at an enormous array of tasks across practically every technical field and industry imaginable. The possibilities are limited only by our imaginations.

Let's brainstorm and get excited about a few more frontiers ripe for change...

Problem Solving and Optimization

From logistics to scheduling, quantum algorithms provide exponential speedups to optimization challenges with countless variables.

Modeling traffic patterns to minimize congestion? Finding the absolute most efficient warehouse design? Perfecting airline crew schedules to lower costs? Quantum crunches options instantly.

Weather Forecasting and Climate Modeling

Today's weather predictions struggle modeling complex atmospheric interactions across massive datasets. Quantum simulation promises exponentially more accurate daily forecasts and long-term climate projections.

Finally - a perfect picnic day based on quantum atmospheric analytics! And hopefully better preparedness for intensifying climate impacts.

Big Data Mining and Analytics

Finding meaningful patterns and correlations in massive messy datasets presents immense computing challenges today. But quantum techniques like Grover's algorithm excel at filtering signal from noise.

Governments to corporations could gain valuable insights from astronomical data volumes. Grover helps find your lost digital keys in endless haystacks of 1s and 0s!

Database Search and Query

Relatedly, quantum search algorithms massively outperform classical methods for querying and retrieving info from huge databases and data lakes quickly.

Goodbye scrolling through pages of search results! Quantum computers produce the perfect personalized answer almost instantly. Siri and Alexa better up their game.

Information Security

Beyond breaking encryption, quantum computers' pattern hunting abilities are also ideal for analyzing malware, insider threats, fraud detection and cyberattack prevention - crunching subtle signals in vast data.

Quantum machine learning assists next-gen cybersecurity defenses by finding anomalies human programmers can miss. Take that hackers!

Logistics and Transportation

Coordinating routes, fleets, traffic, and shipments with endless variables is an epic quantum optimization opportunity. Startups like Quantinuum already offer commercial solutions.

Imagine Amazon deliveries plotted quantum-style - maximally efficient routing that minimizes costs and environmental impacts. A sustainable logistics revolution!

Healthcare Advances

From drug discovery to optimizing hospital flows to uncovering disease risk factors in genome data, quantum stands to massively enhance health outcomes while lowering costs.

Quantum-designed precision medicines tailored to individuals' DNA? Yes please! A quantum-powered push toward preventative, personalized medicine.

Industrial Efficiency

Complex processes from chip fabrication to food production are primed for quantum efficiency boosts. Finding faults faster, reducing waste, and optimizing supply chains through quantum machine learning and simulation.

Lower costs, improved quality, and greater sustainability. It's a quantum industrial revolution! Even quantum computers will design better quantum computers.

Alright, let's catch our breath here because I'm sure your mind is reeling thinking about how quantum computing could transform practically every technical field. Deep exhale...

The key takeaway is that we are just scratching the surface of possibilities. As with past breakthroughs like electricity or silicon chips, applications often emerge over decades of refining a radical new technology.

Patience and persistence will unlock quantum capabilities today considered miracles. But by learning foundations now, you'll be ready to ride the quantum wave as it continues rising!

So although aspects still seem impossibly futuristic, maintain faith in science's ability to gradually tame and harness quantum's massive latent potential through collective discovery. What an exciting time to be alive!

Chapter 6 - The Future of Quantum Computing

Obstacles to Overcome

Alright, we've covered incredible quantum progress unlocking new computing paradigms. But also don't forget - huge hurdles remain to realize the full vision of fault tolerant, universal quantum machines. This rocket ship is still undergoing some engineering tuning before we blast off to Planet Quantum!

But the key is that none of the remaining challenges appear fundamentally insurmountable given enough R&D resources and time. The laws of physics aren't conspiring against us (yet). Patience and persistence will get us to the promised land!

Now that we understand quantum's massive latent potential, it's worth reviewing remaining roadblocks so we can cheer on researchers as they overcome obstacles one by one. Let's run through some of the biggies:

Qubit Stability

By far the biggest challenge is enhancing qubit stability and resilience through improved error correction and fault tolerance. Progress is steadily being made, but we need qubits to maintain their probability superpositions much longer to achieve commercial viability.

Essentially, we need to coax finicky quantum states into cooperating long enough for meaningful computations before noise and interference makes them decohere into classical randomness. It's the fickle physics of fragile probabilities!

But from exotic materials like diamond vacancies to topological anyon models, researchers are exploring radically novel approaches to sustain qubit states. And existing systems are becoming less noise prone through refined engineering of everything from control electronics to literal soundproofing against sonic vibrations. Steady progress continues!

Interconnect Bottlenecks

Another major obstacle is wiring together large grids of logically interconnected qubits to create a functional quantum processor architecture. Enabling smooth controllable coordination between quantum units remains key.

Teams are already testing creative solutions like photonic interconnects versus traditional metal wires to allow robust qubit communication with minimal crosstalk errors. Again, today's challenges will likely inspire tomorrow's breakthroughs!

The Software Stack

Beyond hardware, robust software infrastructure is needed so users can program quantum routines and apps without worrying about the finicky physics underneath. Tools for accessing quantum capabilities need to become as easy as coding a website!

That's why researchers are focused on building full stack solutions like quantum operating systems, hybrid programming frameworks, and algorithm libraries to abstract away the exotic physics and make this powerful technology accessible to entrepreneurs and business users.

Killer Apps Missing

While the potential for quantum advantage spans industries, definitive killer applications demonstrating unambiguous commercial utility above all alternatives remain elusive. The business proposition, while promising, is still hazy.

But again, as hardware and software matures, undeniable use cases will emerge just as apps like email and search became obviously indispensable on classical computers after initial tinkering. The quantum app store still has some stocking up to do!

Alright let's catch our breath here because I'm sure your mind is spinning thinking about fault tolerant machines that remain over the quantum horizon. But take comfort knowing each incremental breakthrough brings us closer to exotic applications!

The key is that the fundamental physics indicates universal quantum computing IS possible - we just need to engineer systems cleverly to coax qubits into cooperating. And financial incentives for progress only accelerate annually.

So despite hurdles ahead, the gauntlet has been thrown down to tame quantum's endless potential through science and engineering. With patience and persistence, a dazzling quantum-powered future awaits!

When Will Useful Quantum Be Real?

Alright, after reviewing the remaining hurdles, the million dollar question is - when can we expect quantum computers to transition from finicky scientific curiosities into mainstream commercial engines powering progress across sectors? Let's gaze into the quantum crystal ball!

Opinions vary wildly on timelines even among experts. Some borderline irresponsible hype suggests quantum will redefine computing practically overnight. Meanwhile, others argue general purpose quantum may remain decades away.

The prudent prognosis likely rests somewhere between the extremes...

Quantum In 5 Years?

Starting with the optimistic perspective, some research leaders and companies insist commercially relevant quantum capacities could emerge in just the next 5 years.

They cite rapid hardware and software advances allowing niche applications like optimizing airline scheduling or plotting warehouse layouts to already run effectively on today's noisy but powerful superconducting quantum processors.

If exponential progress continues, universal error-corrected quantum machines may become viable much sooner than previously thought. Exciting times ahead!

Quantum In 10+ Years?

Meanwhile, other experts caution general quantum usefulness may still be 10 years away or more. Their skepticism stems from just how astronomically difficult engineering robust fault tolerance and error correction mechanisms remains.

Plus, definitive killer apps that justify quantum's additional complexity and cost over improving classical techniques remain elusive. The business case is still uncertain, causing hesitance.

They argue today's accomplishments primarily represent scientific progress, not technological maturity. We should temper expectations and acknowledge the remaining challenges ahead.

The Prudent Path

Of course, the prudent path likely involves embracing the middle ground - being optimistic about continuous incremental progress toward quantum readiness while acknowledging paradigm-shifting capabilities are still years away rather than months.

A balanced perspective combines hope with patience, envisioning exponential change while understanding foundational advances remain. This fuse drives progress while avoiding dangerous hype bubbles.

The next 5 years should continue seeing quantum productivity gains and niche optimizations, even if general quantum dominance remains on the horizon. Useful yet limited capacities first, then exponential transformations to follow!

Alright, let's catch our breath here before getting too fixated on predicting the arrival date of techno-rapture down to the year. Whether 5 or 15 years, profound change looms!

The key is maintaining long-term optimism while focusing short term on the incremental breakthroughs accumulating over this decade across hardware, software, applications and education.

With so many brilliant teams pouring passion into the quantum frontier, we should expect continuity of rapid progress regardless of any single inflection point. The future is quantum!

The Coming Quantum Era

Alright, we've covered the science and systems behind this coming computing revolution. Now let's gaze into the quantum crystal ball to imagine what a world powered by these technologies might actually look and feel like!

How will business, government, academia and more adapt when quantum capabilities emerge from physics labs into the mainstream? Let's speculate on impacts across sectors...

Next-Gen Quantum Businesses

The business models of tomorrow will take advantage of quantum computing abilities we can scarcely even comprehend today. Visionary entrepreneurs will find radically new opportunities.

Disruptive quantum startups will emerge overnight and scale exponentially on the backs of superior optimization, prediction and generative creativity unfettered by conventional constraints. It's quantum entrepreneurship baby!

Meanwhile, established companies will either adapt or perish as quantum reshapes competitive landscapes. Survival necessitates upgrading data and crypto defenses while finding ways to harness quantum analytics.

Quantum skunkworks projects will emerge within all major organizations. Quantum readiness becomes the new imperative across industries.

Economic Ripple Effects

Widespread quantum abilities will ripple outwards as a rising tide that lifts all ships. Logistics, finance, agriculture and more will upgrade thanks to quantum-enhanced efficiency, discovery and sustainability. Standards of living improve for millions.

Of course, with great change comes disruption. Entire professions and business models will prove vulnerable to automation or obsolescence. Responsible perspectives must guide transitions and ensure broadly shared prosperity.

New Policy Needs

Governments too must get quantum ready - everything from trade standards to data regulations will need astute revamping for a quantum economy. Outdated frameworks require modernization.

Thorny policy domains include:

- Updating cryptography laws

- Fair access to quantum resources

- Limiting computational threats

- Antitrust constraints on quantum players

- Ethics safeguards on generative AI

- The policy challenges are immense yet exciting. With wisdom and foresight, we can steer progress for the common welfare.

Rethinking Cybersecurity and Privacy

At the individual level, principles of data security and privacy may transform in unexpected ways thanks to quantum capacities for pattern discernment exceeding human comprehension.

Total secrecy becomes unrealistic when any encrypted data gets cracked instantly. Brute obscurity is no longer protection. What constitutes ethical data usage? A complex rethink awaits.

At the same time, quantum AI assistants may become so attuned to our preferences that true individual customization replaces the need for mass data collection. Bespoke beats big data through prediction accuracy!

Academic Revolutions

Across academia, quantum will spur revolutions in scientific disciplines from physics to chemistry as we refine our understanding of simulated realities. Even philosophy and ethics will need rethinking.

A new generation of quantum-savvy graduates will become highly coveted. Nations will race to build the educational infrastructure required to train quantum-fluent visionaries.

Tangible Transformations

On a visible level, quantum technologies will gradually weave themselves into the fabric of everyday life - from smarter devices to revolutionary medicines to hyper-efficient infrastructure and transportation. Changes compound!

Yet the deepest impacts may remain opaque, woven into prediction algorithms and invisible discoveries only made possible by transcending conventional computing constraints. Quantum possibilities permeate life stealthily.

Alright, let's pause our quantum future gazing here before our imaginations escape too far into science fiction territory!

The core idea is that quantum computing will spur change touching all areas of business, government, science, technology and society given the versatility of its exponential speedups. But the specifics remain thrillingly unclear!

Rather than fixating on point predictions, it's best to nurture our mental flexibility and readiness to understand novel breakthroughs as they emerge. The future remains a moving target.

So let's continue cultivating conceptual agility and a learning mindset as we stand on the cusp of this computing revolution. With vision and values, the quantum age can uplift humanity across the board!

Quantum Risks and Regulations

Alright, let's wrap up our quantum tour by discussing responsible development - appreciating risks alongside opportunities as we build this brave new computing era. With great computational power comes great responsibility!

The fact is, any powerful technology brings associated perils if wielded recklessly - from nuclear energy to genetic engineering. But prudent perspectives can minimize downsides while maximizing upside.

As quantum matures, we must proactively envision potential hazards across security, economics, ethics and more to guide wise policies and norms that steer emerging capabilities toward human betterment. Fortune favors the prepared!

Let's highlight a few key domains for long-term thinking:

Information Security

Encryption risks are already covered - but more broadly, quantum hacking could compromise systems digitally coordinating everything from power grids to banking networks. Vulnerabilities require extensive remediation.

New quantum-resistant encryption methods help. But cyber risks expand as hacking itself relies more on quantum techniques. We may need "quantum police" to enforce regulations and track down digital scofflaws.

Economic Balance

The benefits of exponentially faster commercial computing must be weighed against transitions that could displace millions of jobs. And early quantum access among financial elites could further concentrate economic divides.

Thoughtful policy can spread advantages broadly through job retraining programs, public quantum resources, and keeping critical industries competitive. Quantum for all - guided properly, this rising tide can lift all ships!

Ethics and Laws

Beyond direct harms, unintended second order effects also warrant consideration as quantum abilities ripple across society. Issues like data privacy, liability around AI/algorithms and intellectual property require nuanced rethinking.

By getting ahead of complex issues like bias in algorithmic systems and establishing sensible international norms, we steer progress responsibly. Ethics run at light speed!

Guiding Light of Science

Moving forward, academia and civil society help provide key wisdom around priorities, risks, and social needs as quantum abilities grow. Science itself guides science - maintaining dialog around ethics and opportunities.

Transparent, thoughtful development open to input brings out the best in transformational technologies by grounding progress in humanitarian values and genuine understanding. We're all in this together!

I know thinking through longer-term complexities alongside exciting possibilities requires deep reflection! But discussing prudent development puts us on the optimal path.

The core takeaway is that by proactively envisioning challenges along with incredible benefits, we thoughtfully steer this computing revolution toward uplifting humanity to new heights of flourishing and comprehension.

But it requires sustained ethical vigilance - not just what is possible, but what should be done. With conscientiousness and compassion, a radiant quantum future awaits!

Conclusion

Quantum Revelation Recap

Whew, we did it friends! After an epic quantum quest spanning countless mind-bending concepts, we've finally reached the superposition of endings and beginnings that is the conclusion. Let's revel in all we now comprehend about this coming computing revolution!

When we began, quantum computing seemed shrouded behind an impenetrable probability waveform of mystery and jargon. But through our journey together, we collapsed that uncertainty into crystal clarity and understanding!

Let's breeze through a highlight reel of all the quantum goodness we now grock:

Qubit Basics

We learned qubits are the crazy quantum basic units enabling weird properties like superposition and entanglement that allow exponentially parallel calculations. Their powers exceed binary bits exponentially!

Quantum Algorithms

We explored exotic quantum algorithms like Shor's and Grover's that achieve astonishing speedups on tasks from encryption cracking to searching huge data. The quantum app store overflows with possibility!

Inside Quantum Machines

We demystified what makes quantum computers tick, from delicate dance of superconducting qubits to maintaining quantum coherence in near absolute zero environments. Mad science!

Key Quantum Players

We surveyed quantum champs like IBM, Google, Microsoft and emerging contenders like startup IonQ. With so much brainpower on the case, the quantum future looks bright!

Quantum Revolution Ahead

We envisioned incredible applications for optimization, chemistry, AI and more. And a world where quantum capabilities transform everything from medicine to security. The possibilities span to the horizon!

Obstacles and Ethics

We acknowledged remaining challenges like error correction, while emphasizing ethical progress. With wisdom and compassion, we steer quantum towards uplifting humanity holistically.

Alright, let's pause our concluding reflections because I'm sure your mind still overflows attempting to digest so many reality-warping information quanta! Don't worry, integration takes time.

The key takeaway is that you now possess the foundational quantum knowledge to appreciate coming headlines and breakthroughs. You hold the kernels containing this computing revolution's immense possibilities!

Understanding will continue unfolding, but you've built the scaffolding. We stand at the genesis of an computing era that promises marvels. But society must shape it for good.

So remain curious, open-minded and thoughtful as we progress into this quantum-strewn future! Be an active learner, creative visionary and compassionate voice. This transformation requires our collective guidance.

The human story echoes down the ages - may we write the next uplifting chapter together!

Where to Go From Here?

Alright my friend, you've completed your introductory quantum computing tour de force! But of course, this is only the start of an endless journey into the quantum realm. Let's chat about where to go from here to keep expanding your knowledge. The quantum train has left the station!

Now that you have core understandings in place, you're ready to start diving deeper into the equations, physics, and technical materials that underpin quantum information technologies. An exciting new frontier of knowledge awaits!

Here are some ideas on where to continue your adventures:

Hit The Quantum Textbooks

For foundational physics knowledge, begin working through seminal quantum textbooks like Nielsen & Chuang's "Quantum Computation and Quantum Information" or Preskill's "Quantum Computing and Quantum Information".

These will provide the complete theoretical groundwork if you're keen to geek out on the math and science fundamentals. No skipping the homework with these college-level tomes!

Get Comfortable With Code

If you eventually want hands-on programming experience for building quantum algorithms and applications, start familiarizing yourself with quantum-friendly coding languages like Python and Qiskit.

Hit up some programming tutorials to get comfortable coding so you can transition smoothly into quantum computing languages like Q# once you're ready for practical quantum.

Engage With Online Courses

For more structured quantum learning, enroll in some of the great quantum computing courses now offered on platforms like Coursera and edX.

Programs offered from places like MIT provide engaging video lectures explain core concepts, along with quizzes and projects to accelerate your quantum chops.

Join The Online Community

Engage more with the quantum community on forums like Reddit to exchange ideas with like-minded learners and get exposed to the latest cutting edge news and discussions.

Seeing real-world progress and applications will complement your conceptual knowledge and feed your quantum excitement!

Try Out Quantum Demos

Sign up for access to commercial quantum hardware like IBM's Quantum Experience to run experiments on real quantum processors via the cloud.

Testing quantum circuits yourself and seeing the bizarre probabilistic results firsthand will concretely reinforce the wild quantum properties you learned about!

Attend Local Events

Look for quantum meetups and conferences in your local area that convene experts across business, government, and academia to share knowledge and demo emerging tech.

Immersing yourself in the quantum scene, even as a beginner, will open learning opportunities and relationships that fuel your growth.

Stay Tuned To Breakthroughs

Keep frequent tabs on leading quantum research labs, startups, and companies via quantum newsletters, blogs, websites and more to be clued into the latest headline-making advances as they happen.

This inspires your learning and helps track the exponential acceleration of commercial quantum progress in real-time.

Alright my determined quantum padawan learner, we've come to the close of your initial training! You now have all the resources you need to continue mastering quantum computing in ways that excite you most.

But never stop nurturing your curiosity across this endless quantum landscape. A beginner's mindset opens the most possibilities!

Remember - learning quantum is a marathon, not a sprint. Let your knowledge and skills compound gradually like quantum interest. With patience and passion, you will go far.

Now go grab this world by the qubits! Thanks for letting me be your quantum guide. But this is only the start of an epic adventure. Onward!

Bonus Deep Dive 1: Quantum Supremacy

Brace yourselves friends, it's time to dive into the science fiction sounding yet very real concept of quantum supremacy!

This term gets thrown around a lot these days, but what does it actually mean? Is it just more hype or a true inflection point along the path to bonafide quantum capabilities?

Well, in typical fashion, the reality lies somewhere in between. Quantum supremacy is an important milestone, but it also doesn't mean our quantum dreams have suddenly materialized overnight. As with most things quantum, the details get...complicated.

So strap on your photon packs and let's unpack Schrödinger's box of what quantum supremacy is, why it matters, and where this technological cat currently sits between entangled states of astonishing progress and remaining practical challenges. Fascinating times!

Defining Quantum Supremacy

First, what does quantum supremacy mean? Simply put, it's the tipping point where a quantum computer can complete at least one mathematical task substantially faster than even the beefiest classical supercomputer conceivably could.

Note the "one task" detail - this isn't full on quantum domination over all computing. We're talking more like a quantum parlor trick specifically designed to highlight the exponential parallel processing advantages these machines offer. Their first flash of greatness!

The key is demonstrating unambiguously for the first time that quantum systems have crossed into doing things fundamentally impossible for normal silicon computers, even hypothetical ones. A watershed moment!

This requires meticulously crafting a benchmark problem that perfectly balances being solvable by fragile quantum hardware yet intractable for simulations on conventional devices. The quantum gauntlet is thrown down!

Why Supremacy Matters

Alright, so why does this narrow demonstration matter? Is it just hype and bragging rights?

Well, consider humanity's epic quest to achieve powered flight as an analogy. Those initial 12 second wobbly flights by the Wright brothers didn't immediately enable practical air travel.

But they PROVED the basic concept's viability and cracked open the runway towards rapid progress. The sky was no longer the limit after that first soaring supremacy!

Quantum supremacy is like computing's Kitty Hawk moment - definitively showing breakthrough capabilities are possible through mastering exotic quantum physics. The floodgates suddenly open to explore full potential.

It transforms quantum from far-off theory to an engineering challenge. The titanic ascent towards useful quantum machines begins!

Supremacy also electrifies research focus and funding. When tangible speedup gets demonstrated instead of just hypothesized, bright minds and big money flood in to chase the vision.

Imagine the trajectory of aviation if the Wright brothers crashed. But their short crafts stayed aloft just long enough to unleash our dreams of the sky. Quantum supremacy opens the same window to awe-inspiring possibilities!

Of course, like those early flimsy flyers, don't expect to be booking useful quantum transatlantic flights anytime soon. But supremacy signals the technology now has enough lift to eventually soar. Metaphorically speaking!

Journey to Supremacy

Alright, let's ground this a bit by reviewing the winding journey leading to quantum supremacy thus far...

The concept was formally proposed back in 2012 by Caltech's John Preskill, who framed these benchmarks as motivating milestones to guide research. Clever guy!

Teams at institutions like Google immediately began work to design and build suitable test cases. This stomach-churning quantum rollercoaster ride was about to leave the station!

Years of meticulous hardware construction and fine-tuning calculation techniques followed. All leading towards crafting the perfect supremacy snapshot that balanced highlighting quantum's promise while minimizing its impracticalities.

The finish line was almost crossed in 2017 when researchers believed they achieved the milestone using ultracold atoms. But a refined classical algorithm burst their bubble by handling the proposed problem reasonably after all. Back to the drawing board!

But Google preserved, and by 2019 they made headlines worldwide for finally attaining undisputed quantum supremacy. This was the real deal!

They built a 53 qubit quantum processor named Sycamore, which sampled random quantum circuits to produce a result in 200 seconds. An impressive feat.

But here's the kicker - they estimated that even using all of Summit, the world's mightiest supercomputer, this specific sampling would take 10,000 years! A jaw-dropping speedup. Quantum's potential was finally proven!

Even John Preskill himself pronounced Google's achievement as crossing the quantum Rubicon, saying: "They have demonstrated conclusively a quantum computer doing something that a classical computer cannot practically do at all."

A historic quantum sup-air-macy moment! Even skeptics like IBM conceded this passed the cryptography sniff test as a valid demonstration, though they quibbled with the hype around timescales. But the genie was out of the bottle.

This watershed sent waves of excitement and optimism through the quantum community. The era of broad quantum advantage was finally peeking over the horizon!

Alright, let's pause our supremacy celebration here to catch our breath and get perspective. Yes, this milestone matters. But misconceptions abound...let's clarify!

What Supremacy Isn't

First, some common misperceptions. Supremacy does NOT mean:

- Quantum computers now beat classical machines at practically useful applications. We're still years away from that.

- Encryption and Bitcoin are immediately at risk of getting hacked in minutes. Quantum key distribution could still take decades.

- We can now accurately simulate novel molecules or build sci-fi quantum AIs. Stability and engineering challenges remain.

Again, think back to airplanes - Kitty Hawk wasn't the start of cross-Atlantic commercial flights. Supremacy only proves viability, not immediate utility. Milestone, absolutely. But measured expectations are prudent.

It also doesn't fully "prove" quantum computers will ever reach fault tolerant universal status rivaling classical systems. Physics indicate yes, but years of hardware work remain to make that a reality.

In essence, quantum supremacy is a powerful signal of progress but not itself the direct finish line. Much refinement is still needed translating these controlled tests into practical advantage.

It's a gust of quantum wind filling our sails, not arrival at the shore of scalable quantum computing. But it's an invaluable navigation marker showing we're now pointed in the right direction, at least!

Supremacy is also limited to a narrow test case, just like those 12 second flights didn't unlock much air travel utility. We now know quantum can soar, but useful cargo transport remains distant.

So in summary, supremacy doesn't herald sudden quantum omnipotence, but it does confirm this exotic technology has graduated from pure theory into credible engineering territory. The runways are built - now for the long flight ahead!

Alright, with a healthy perspective on supremacy's limits, are you ready to envision the industries this milestone could transform once qunatum machinations mature? Buckle up, we're crystal ball gazing into an awe-inspiring future!

Quantum Springs Eternal on the Supremacy Horizon

For decades, we dreamed of this quantum rocket finally lighting and beginning to lift off the launch pad. And with Google's recent supremacy success, it feels like reality caught up to theory.

The precedent is powerfully set. Yet still the universal quantum horizon beckons even as we celebrate this milestone.

Supremacy opens our minds to credibly ask - just how might our world transform once these systems smoothly and reliably solve useful problems beyond contrived tests? Let's speculate...

Future Computing Itself

Let's start big picture. Once scaling challenges are solved, fault tolerant quantum computers promise to inherit the earth, ushering in computing infrastucture exponentially more powerful than silicon chips. Hello world!

These machines won't wipe classical systems out entirely. Similar to analog persisting alongside digital, niches favoring robustness over quantum's probabilistic delicacy will remain.

But for applications benefiting from incomprehensible multi-dimensional parallelism, quantum will quickly come to dominate processing needs throughout society and business. A hybrid classical-quantum computing ecosystem emerges.

We'll wield quantum machines so advanced they design improved versions of themselves. Sorcery! But with ethical care, this could unlock a utopian age of discovery.

Artificial Intelligence and Machine Learning

With exponentially more "brain" power, quantum machine learning promises revolutions in artificial intelligence - from quantum neural networks that mimic our own mind's uncertain nature to generative creativity exceeding human imagination. And that's just the start!

Training complex quantum models on massive datasets for unprecedented predictive prowess becomes practical. New breeds of quantum learning algorithms tackle tasks like computer vision, language translation, drug discovery and much more with almost unthinkable proficiency.

Sophisticated quantum AI assistants, advisors, creators, and discoverers could almost magically materialize to cooperate with us in everything from science to art to minimizing traffic jams. But wisdom in wielding such tools will be key. With ethics and oversight, quantum AI could propel a true renaissance!

Security and Encryption

Of course, quantum capabilities cut both ways. On one hand, blockchain and cryptography face reckoning from algorithms like Shor's that crack modern encryption schemes with ease. Everything from bank data to state secrets becomes vulnerable.

But quantum principles also allow radically powerful quantum-safe encryption if we're proactive in migrating. With quantum's own tricks, we can stay steps ahead of those who may threaten digital security

- if society makes prudent preparations now for the inflection point supremacy promises.

Financial Systems and Optimization

Quantum computing's data crunching muscle also promises to transform sectors like finance and logistics through unprecedented optimization abilities. Tasks like dynamic portfolio optimization, derivatives pricing, smart contract handling, and supply chain coordination get hypercharged.

Platforms like quantum-boosted algorithmic trading reshape markets through sheer knowledge at scale. But again, wise policy and foresight helps spread advantages broadly. With care, optimization computing can also help tackle humanity's greatest challenges like inequality, sustainability and more by elevating global wisdom.

Healthcare and Chemistry

Revolutionary healthcare treatments, materials, and energy solutions also await discovery through simulated quantum laboratories. Modeling complex molecular and atomic forces with fidelity allows designing future pharmaceuticals, catalysts, and supermaterials atom by atom.

A golden age of renewable energy, programmable matter, and personalized medicine beckons. Even aging itself may eventually yield to controllable chemistry as quantum principles unlock DNA's deepest secrets. A long and vibrant future for all could await.

Alright let's pause our speculative sci-fi pondering here before proclaiming the coming quantum utopia! True potentials remain vague. But supremacy opens portals to barely fathomable possibilities if we tread mindfully.

The Next Leap After Supremacy

So in summary, achieving quantum supremacy reminds us that transforming early science experiments into world-changing technologies requires immense subsequent engineering once basic viability gets proven.

We celebrate seminal moments like first flights, microchips, vaccines, or internet connections - yet the real wealth emerges gradually from then optimizing, scaling, and applying those breakthroughs broadly. The real headline is never the first lab prototype, but rather the society-shifting maturation that follows.

And with quantum, we still have a long road ahead translating fragile but blazingly fast contrived demonstrations into optimized fault-tolerant reliability. We must walk before we run!

But the historic glimpse behind the veil provided by quantum supremacy cannot be overstated either. We envisioned aviation's destiny thanks to that first wobbly flyer, even if useful 747s took decades more. Supremacy echoes the same sense of incipient amazement and focus.

So a balanced posture combining patient optimism with perspective serves us best. Having tasted quantum's potential, we must now thoughtfully steward its co-development with wisdom, ethics, and compassion as guiding lights.

With conscientious shepherding, humanity could stand ready to harness quantum's eventual revolution for the benefit of all peoples. But sound policy and priorities determine if coming transformations uplift or undermine universal flourishing. Our choices matter!

So in closing, quantum supremacy deserves celebration for crossing a seminal scientific milestone. But even more, it warrants reflection on thoughtful collaboration needed next to gently nourish this knowledge into humane technologies exceeding previous limits.

With care, someday quantum computers may not only exceed conventional compute, but help develop our civilization's own collective maturity through compassion. But first, we must walk - with ethics ever guiding our steps into the wide quantum horizon now peeking into view.

Onward, with joy, creativity and consideration! The future awaits.

Attention Quantum Enthusiasts - Your Mission, Should You Choose To Quantum Leap...

Greetings, fellow quantum explorers! Now that you've embarked on this intriguing journey into the realm of quantum computing, I have a humble request to make.

In the spirit of sharing knowledge, it would be truly phenomenal if you could spare a moment to leave a genuine and positive review of your experience with this book. I know, I know - please indulge me for a moment!

You see, your feedback holds more significance than you might imagine. Allocating a brief time to express your positive insights wields remarkable influence.

Positive reviews send a mysterious signal to the intricate algorithms of literature, signifying that the insights on quantum computing presented here are genuinely aiding curious minds (that includes you!).

This signal then amplifies the book's visibility, allowing others to also delve into the world of quantum knowledge. Your review becomes a catalyst for spreading enlightenment.

Moreover, your distinct viewpoint might be the exact push future readers need to embrace the journey into quantum possibilities. Think of it as a recommendation from a trusted colleague, resonating with you, right?

Especially for an independent gem like this quantum book, reviews are of immense help. So if you found resonance and value within these pages, kindly lend your voice to magnify the resonance!

The ultimate goal is to bring these quantum concepts to as many inquisitive minds as possible. Your review propels this noble endeavor.

With heartfelt appreciation, thank you for even contemplating this endeavor.

Onward and upward in the quantum realm, companions.

www.ingramcontent.com/pod-product-compliance
Lightning Source LLC
Chambersburg PA
CBHW071323130726
47996CB00002B/597